T0379403

Building Women Leaders

Building Women Leaders

A Blueprint for Women Thriving in Construction

Gretchen Gagel, PhD

Library of Congress Cataloging-in-Publication Data

Names: Gagel, Gretchen author
Title: Building women leaders : women thriving in construction / Gretchen Gagel.
Description: Hoboken, NJ : Wiley, [2025] | Includes bibliographical references and index.
Identifiers: LCCN 2025002987 (print) | LCCN 2025002988 (ebook) | ISBN 9781394251384 paperback | ISBN 9781394251421 adobe pdf | ISBN 9781394251407 epub
Subjects: LCSH: Leadership in women | Construction industry–Vocational guidance
Classification: LCC HD6054.3 .G33 2025 (print) | LCC HD6054.3 (ebook) | DDC 658.4/09082–dc23/eng/20250321
LC record available at https://lccn.loc.gov/2025002987
LC ebook record available at https://lccn.loc.gov/2025002988

Cover Design: Wiley

Set in 9.5/12.5pt STIXTwoText by Straive, Chennai, India

SKY10101923_040425

This book is dedicated to my two beautiful children, Holden and Regan, who are the joy of my life and the inspiration for all I do in the world; and to the memory of Edgar Schein, my friend and coach, amazing author, academic, consultant, and thinker. You are missed and I think you would be proud.

This book is also dedicated to the millions of people in the construction industry who work tirelessly to build and maintain the physical assets of civilization with humility and grace.

Contents

About the Author

Dr. Gretchen Gagel is passionate about leveraging her successful career as a construction industry leader to continue influencing positive change in the industry she loves. Gretchen serves as a leadership role model via her industry board and committee positions, and supports the development of leaders through her leadership programs and executive coaching. Gretchen is also the Founder of Women Thriving in Construction: A Global Institute (www.womenthrivinginconstruction.org). Gretchen is a sought-after speaker on a variety of topics.

Gretchen is grateful to have wonderful friends and family, including her two beautiful grown children, and is a passionate golfer and enthusiastic world traveler.

You can find out more about Gretchen at gretchengagel.com or connect with her on LinkedIn.

Foreword by Janice L. Tuchman

I've been observing Gretchen Gagel for more than 25 years—basking in her big smile, benefitting from her wonderful connections, and marveling at her productivity and leadership. Now I find out that she also excels at storytelling! The following book offers both personal stories and stories Gretchen relays from her collection of accomplished friends and colleagues. It offers a wide array of ideas, tips, and resources for women who lead or aspire to lead in the construction industry, as well as those who work with women leaders.

Gretchen and I have a mutual admiration for each other's broad networks in the industry, and it's been a two-way street on mentoring. We met as speakers at a West Coast construction owner's association meeting. I was daunted to follow her at the podium. She is such a natural presenter—so poised and knowledgeable. No written-out speeches for her. It made me realize I should step up my game. And her outfit! A new generation of women now are taking their cue to business style from Kamala Harris—but it was Gretchen Gagel who inspired my wardrobe shift to presenting in a pantsuit.

Gretchen gives readers permission up front to "snack" on this book—to poke through and consume it either all at once from beginning to end (as the engineer readers almost surely will do) or instead to jump to a chapter that sparks interest, read some nuggets, and then jump ahead or back to a different section of interest on a particular day.

What did I read first? When Gretchen sent the final draft to fellow foreword-writer Hugh Rice and me, she told us the chapters where she quoted us. Curiosity sent me first to Chapter 22, where both of us make an appearance—among her interviews with 100-plus women and men. The discussion is about leading not just our own teams and organizations, but also working to help shape the future of construction. Throughout the book, Gretchen gives lots of ideas about how that can be done. For example, she advocates getting involved in industry groups, raising your hand for a committee or job, and taking action about an issue you feel passionate about.

One of the issues both Gretchen and I feel passionate about is how infrastructure can help underserved communities. In 2015, at the invitation of its brilliant and charismatic CEO Avery Bang, we both joined the board of directors at the nonprofit Bridges to Prosperity. B2P builds footbridges to connect remote communities cut off by raging rivers from schools, jobs, and health care. It now works actively in East Africa where the need is great, and the population density is high. Research has shown that adding a footbridge can raise a community's income by up to 30%—and particularly, it helps get girls to school. About five years into our terms, Gretchen was a natural to become the next board chair. She had led consulting companies. She studied leadership theory for her doctorate. Her list of positive impacts was long. When her term was up, a journalist—me—was a less-obvious choice. But I raised my hand—with Gretchen's encouragement and advice—and found my collaboration and problem-solving skills did help the group during a time of transition.

Back in Chapter 1, I found Gretchen's personal and related stories about sexual harassment, bullying, and unconscious bias both compelling and frank. She lays out the issues, explains how personal experiences frame her views, and analyzes how such situations make it hard to achieve an inclusive workplace. One of the stories made me think of a small printed sign I had on my wall at Engineering News-Record in the late 70s. It said, "Women have to do twice as well as men to be thought half as good. Fortunately, this is not hard." I relished how peeved the little sign made my male manager at the time. But Gretchen's chapter made me think through the experiences that led to the need for my show of bravado.

Throughout, Gretchen lays out her vision for construction as an inclusive industry and her belief that this vision is achievable. That's why she gravitated to the title: *Building Women Leaders: a Blueprint for Women Thriving in Construction*. Thank you, Gretchen, for sharing your vision and for laying out a critical path to that future.

Janice L. Tuchman
ENR Editor-in-Chief Emeritus and Events Consultant
Member, National Academy of Construction

Foreword by Hugh L. Rice

This is a book on leadership. But it is not a typical high-level overview of leadership theory. It is rather a "How To" book. It is a practical step-by-step guide to the discipline of leadership and management. Also, this book is the personal story of Gretchen Gagel's journey through her education and career choices and the discoveries she made about how to be a leader. And not just a leader, but a leader in the male-dominated construction industry. And finally, this book is a call to arms and guide for women who want to become leaders and to do it in the patriarchal world of construction. Leadership is important. Leadership is critical. We all know organizations where the person at the top is the key to success or failure. Leadership matters.

Women make up roughly 50% of the U.S. labor force. But the work force of the construction industry is between 5 and 10% female, depending on how you count. And most of those are in office/administrative roles. The construction industry is ignoring half of the American labor force—and it is the smart half! Gretchen describes in detail her experience with bias, prejudice, and condescension while working in the male-dominant culture of the manufacturing and construction industries. She succeeded through hard work and force of personality and is passionate about helping other women succeed also.

It has been well documented over the past couple of decades that diversity works. Diverse groups inevitably produce better outcomes than homogeneous groups. It is true in committees, business units, and organizations of all types. Since the United States has a multicultural, multi-gender, and multiethnic work force, no country is better equipped to capitalize on diversity in the work force. In fact, given the labor shortages facing construction and other industries, having a diverse work force is mandatory. There are simply not enough white males, or males of any type, to meet the needs of the construction industry. Women are the obvious underutilized segment of the labor force who can fill the void.

One thing you will notice about this book is that Gretchen has a unique personal perspective on being inclusive and breaking into a new culture. She has 40 years of lived experiences that she shares freely. Gretchen is a student of leadership and management in construction, both academically and as a practitioner. She learns from others and internalizes everything. She never forgets anything that she hears, reads, or experiences, or anyone she meets. Since Gretchen is a consummate networker and relationship builder, we all have the benefit of what she has learned from hundreds of people in the industry.

Reading this book reinforces a belief that I have had for many years—the world would be a kinder, gentler, safer, and more prosperous place if all heads of state on the planet were women. Now I would add that at least half of the CEOs of construction firms should be women. Someday it will be so.

Hugh L. Rice
Chair Emeritas, FMI Corporation
Member, National Academy of Construction

Preface

In 1981, at the age of 17, I told my father I wanted to go to engineering school. His response? "Girls don't study engineering." Dad wanted me to get my MRS degree to find a nice man who would take care of me as he had taken care of my mother. When I turned down my acceptance to Harvard to attend engineering school at Southern Methodist University, dad disowned me, and we did not speak for quite some time. In the end, he was proud of my career, but it was a difficult time for me. I believe this experience fueled my desire to succeed and created a strong awareness of the unconscious bias many of us face.

I spent the next 35 years "succeeding" as the only female engineering intern at Lone Star Gas in 1983; the first female operations manager at Ralston Purina in the United States with 62 male peers in 1986; the second female consultant, third female director/shareholder, and first female business unit managing director at FMI; the first female CEO/president of Continuum Advisory Group; the first female chair of Brinkman Construction; and in starting up my own companies in both the United States and Australia. Rarely did I think of myself as a role model paving the way for other women in the construction industry. I was just doing the things I loved to do. Was it hard? Yes. I am confident my father's words to me were often ringing in my ears as I strove to succeed.

My 2021 induction into the US National Academy of Construction was a liberating moment for me. I realized that I had nothing left to prove to my father or anyone else in the construction industry who may have doubted my abilities because of my gender. I also realized that my time away from the construction industry from 2006 to 2012—in part as CEO/president of The Women's Foundation of Colorado (WFCO)—was an incredible opportunity to build a network of women friends and mentors that I would not have experienced had I not left the construction industry. This, combined with my experience of starting a women's leadership development program for the Australian Pipelines and Gas Association (APGA) in 2019, fueled my passion for helping other women thrive in the construction industry. In 2022, I started to think big. Maybe I could impact 10,000 or 100,000 women in the construction industry, and in turn help shape a more inclusive industry. Maybe I could build a movement.

I also realized during my PhD studies that most of the business and personal development authors I admired were men—Jim Collins, Michael Beer, Stephen R. Covey, Peter Senge, and Peter Drucker. Why weren't their more top-selling business books by and for women? Friend, speaker and author Holly Ransom was inspired to write her book *The Leading Edge* based on her leadership research during her time as a Fulbright Scholar at Harvard. *"There were two things that really struck me. One is it's a pretty one-dimensional set of leaders that we hold up as exemplars. They're overwhelmingly people that are from a military context, maybe from an industrial age business model, the Jack Welsh's of the world, the Silicon Valley CEO types. There was a real lack of the stories of women. There was a real lack of cultural diversity. There was a real lack of generational diversity. This idea that leadership can take many forms and can come from anywhere didn't seem to be as well reflected for me in the leadership literature"* (Ransom, 2021; Holly Ransom, speaking to Gagel, 2022 Greatness Podcast).

I feel fortunate that Kalli Schultea at Wiley Publishing began following me on LinkedIn and said, "I think this person has something to say about women leading in construction." When Kalli approached me about writing this book, I realized that yes, I do have something to say on this topic. I love the construction industry and have worked on systemic change within the industry my entire career. My 2018 PhD in leadership, organization culture, and organization change and agility—along with four decades of observing thousands of leaders and my own leadership experiences—provided the expertise, and my passion for helping women provided the motivation.

The purpose of this book evolved over time. Initially, I wanted to inspire women to want to enter, stay, and thrive as leaders in the construction industry. Then, I realized I also wanted to help people who do not identify as women understand how to help women enter, stay, and thrive as leaders in the construction industry. Thank you to the people who identify as male who are reading this book. I hope this book helps you thrive as a leader and understand how to help the women around you thrive as well. As I began writing, I realized there was also an autobiographical purpose to capture my experiences, my stories, my successes and failures, and the lessons I have learned.

I wrote this book from the perspective of a person who identifies as female and Caucasian in the construction industry, but many of the ideas I share apply to other underrepresented communities in a variety of industries. I have included the voices of over 100 additional people I admire, people who have appeared as guests on my Greatness Podcast, and the women and men I interviewed specifically for this book. Their stories bring the ideas I share to life.

Figure 1 Dr. Edgar Schein and Dr. Gretchen Gagel, June 27, 2017, Palo Alto, California.

I mention the wonderful Edgar Schein and his son Peter often in this book. I discovered Edgar and his work during my PhD studies—an amazing MIT academic, consultant, and author of over 40 books on leadership and organization culture. I worked up the courage to reach out to Edgar via email to tell him how much I admired his work, and he invited me to visit him, which I did on June 27, 2017 (Figure 1). In early 2021, I reached out to Edgar for advice about my complicated life and he became my coach for a year before sadly passing away at the age of 94 in January 2023. I am grateful for the many gifts Edgar gave me as he shared his wisdom, and I am excited to have the opportunity to share much of his knowledge and my learnings from him in this book.

I hope you enjoy reading this book as much as I enjoyed writing it. Thank you for including me in your leadership journey.

References

Gagel, G. and Ransom, H. (2022). *Holly Ransom discusses her book The Leading Edge* [Podcast]. 30 September. Available at: https://open.spotify.com/episode/ 0M5G6yRwqvbBlKa302wmKb [Accessed: 1 May 2024].

Ransom, H. (2021). *The leading edge: Dream big, spark change and become the leader the world needs you to be*. North Sydney: Penguin Random House Australia.

Acknowledgments

I am grateful for the many people who helped me complete this book, including the friends and family who supported my writing journey. My friend Jill Tietjen, fellow member of the National Academy of Construction and author of several books, read the initial draft of this book and provided many insightful comments—thank you! Fellow Wiley authors such as Gabrielle Dolan and Rania Anderson also gave me wonderful guidance along the way. Many, many people have taken the time to be interviewed or appear on my Greatness Podcast. Thank you for your contributions to the knowledge contained in this book.

Thank you Ha P. Tran for your work on many of the images included in this book. Thank you as well to the entire Wiley Publishing Team, including Nandhini Karuppiah, Hariharan Jayamoorthy, and Jeevaghan Devapal. Of course, it all started with Kalli Schultea—thank you for finding me and giving me voice.

Testimonials

"Observing Gretchen's leadership as Chair of international non-profit Bridges to Prosperity and experiencing our friendship, I am confident that this book will benefit many. As a leader, Gretchen is not afraid of making the tough decisions, is adept at handling conflict, and works hard to ensure other's voices are heard. I am thrilled that she is sharing her leadership insights and her passion for helping women thrive in the construction industry with the world."

—Carole Bionda, Owner, CLBionda Consulting, and retired Vice President, Nova Group, Inc.

"In the six short years that Gretchen has been in Australia, it has been evident that she is passionate about the construction industry and driven to help women succeed. To reestablish yourself in our industry as a woman in her fifties took grit and determination. Gretchen's genuineness is apparent, and I know many will benefit from her knowledge of the elements of great leadership and how women can succeed in our industry."

—Meg Redwin, General Counsel and Executive Director at Multiplex

"Gretchen is a force of nature and true champion for change in the construction industry. Along with our current ACA President Annabel Crookes, and the many women who serve on our board, she is paving the way for the women leaders we need if this industry to be successful. This book is an important resource for us to understand the role we all can (and must) play in creating a more diverse and sustainable industry."

—Jon Davies, CEO, Australian Constructors Association

"Working with Gretchen over the past thirty years has been a true pleasure. As a leader, she excels at maximizing the potential of individuals, teams, and organizations within our industry. Gretchen embraces her role as a **positive disruptor**, motivating us to reach new heights. Her dedication to empowering women in our field is particularly noteworthy, reflecting her commitment to fostering growth and innovation for all."

—Michael Mayra, retired Construction Group Manager, General Motors

Introduction

I learned two things from the experience of my father telling me I could not attend engineering school because "girls don't do engineering." I learned that there are people in the world who have opinions about what other people can or cannot be because of unconscious bias, and I learned that I have the confidence to make what I believe are the best decisions for my happiness and success. This experience laid the groundwork for my passion for ensuring that all people are able to pursue their dreams regardless of their gender identity, skin color, sexual orientation, socioeconomic background, or any other unique characteristic. We all belong in the construction industry, and the construction industry needs all of us.

We must attract women and other underrepresented groups to our industry—not because it is "nice," but because it is a business imperative for these reasons:

1) **Our ongoing talent shortage:** This is an issue the construction industry has faced for decades. As an example, in 2023, the US construction industry faced a shortage of roughly 650,000 people (LaRocco and Goldberg, 2023).
2) **The benefits of diverse thinking:** Diverse thinking brings about better team performance. For example, research demonstrates that teams with women on them outperform teams without women in solving problems ('What makes a high-performing team? The answer may surprise you,' 2017).

> As Andre Noonan, Chief Operating Officer of ACCIONA Australia and New Zealand shares, *"It is a great industry that I love, and it really irks me that ninety percent of the people within the industry are from fifty percent of the population. It's just nonsensical. The other fifty percent are missing out on what is so great."*
>
> (Andre Noonan, speaking to Gagel, 2022 Greatness Podcast)

The intent of this book is to: (1) help us examine the lived experience of all people—especially women—in the construction and related industries such as engineering, energy, and mining; (2) help all people thrive as leaders by exploring what I believe are the critical elements of great leadership; and (3) help us build a more inclusive construction industry. This book is a culmination of my 40 years in the construction industry as a leader and developer of leaders and my study of leadership theory during my PhD. Many of my life stories are included in these pages, along with the stories of over 100 incredible thought leaders who have appeared as guests on my Greatness Podcast or whom I have interviewed for this book.

I asked every person I interviewed to share their thoughts on what skills had contributed to the success of the women they admire in the construction industry, and several themes arose. The first is that you must be technically competent in your area of expertise (such as engineering, project management, accounting, marketing, or business development), as this is the ticket to the dance. You may not know everything in your field; the point is that you have knowledge and expertise, and you are working hard to continue learning your craft. The people I interviewed also believed that what truly propelled women's success in the construction industry was a belief in themselves as leaders and others believing in them (sometimes even more than they did themselves) and all of those really "hard skills" that have gained the misnomer of the "soft skills" (the ability to communicate, build relationships, garner trust, and resolve disputes, for example). These are the topics I have focused on in this book.

> As construction industry leader Donna McDowall shares, "*Know your value and be sure of yourself. Be okay being the diverse you and recognize it takes every skill set to make a project; and don't give up. It is a good industry.*"
>
> (McDowall, 2023)

The book's content follows a framework of leadership I have labeled the **"ME, WE, and US"** of leadership, first focusing on the "ME" of Grounded Self-Leadership, then the "WE" of building strong individual relationships, and finally, the "US" of effectively leading teams, organizations, and even the construction industry as a whole.

ME

While the intent of this book is to be inspiring and future-focused by exploring how women thrive in the construction industry and how our allies support us, I felt it would be remiss not to first reflect upon the lived experience of women in the construction industry. Part 1 of "ME," **The Lived Experience of the Construction Industry**, touches on this topic, including the presence of unconscious bias in our industry and the realities of our dominant culture.

Part 2 of "ME" is an exploration of self-leadership. I have labeled this the **Grounded Self-Leadership Framework** because I believe being an authentic leader must come from a place of strong self-identity and knowledge of our uniqueness and that this is "grounded" in our purpose and values. The topics covered in Part 2 include:

- Defining Leadership and Your Leadership Aspirations – An exploration of what leadership means to you, how leadership differs from management, and your leadership aspirations.
- Living Your Life Purpose and Values – An exploration of why you exist in the world and your personal values as the foundation you stand upon as a leader.
- Feeling Courageous and Confident with Humility – An exploration of the courage, confidence, and humility required to lead and influence, especially as women in male-dominated industries.
- Thinking Critically, Learning, and Reflecting – An exploration of how we effectively utilize our brains to think and learn, as well as rest to fuel innovation and creativity.
- Telling Your Leadership Story – An exploration of the importance of being able to tell the story of our leadership brand to others and have a voice.

These elements form who you are as an authentic, unique, and grounded leader.

WE

The "WE" portion of the book explores how we build strong one-on-one relationships and leverage the help of others to thrive as leaders. Part 3 of "WE," **Your Supporting Cast**, focuses on the amazing people who guide and support us (that I refer to in my life as "Team Gretchen"). This includes individuals—mentors, sponsors, coaches, and allies—and our networks both inside and outside of the industry.

Part 4 of "WE" focuses on **Building Relationships** with people, which is the next critical step in leadership. The topics covered in Part 4 include:

- Being Present, Listening, and Asking Great Questions – An exploration of how being in strong relationships begins with the ability to be fully present with someone, including the ability to listen and ask great questions.
- Utilizing Power to Influence – An exploration of our sources of power and how to use an appropriate level of power to influence others.
- Leveraging Emotional Intelligence – An exploration of how we experience and control our own emotions to avoid triggered reactions, and how we have empathy for others' emotions.
- Building Trust and Psychological Safety – An exploration of how we effectively build the trust and psychological safety necessary for meaningful relationships.
- Collaboration, Communication, and Conflict Resolution – An exploration of what I call the "Three Big C's" that are critical to our relationships with others.

- Clarifying Expectations and Providing Feedback – An exploration of the skills that are critical in helping others understand what a "good job" looks like and coaching them to success.

These elements provide the skills necessary to form strong relationships with those you lead and influence.

US

The third portion of the book broadens our viewpoint to leading and influencing collective groups of people. Part 5 of "US," **Leading Teams and Organizations,** covers these important topics:

- Defining Purpose, Goals, and Objectives – An exploration of how teams and organizations align on why they exist and what they are trying to achieve.
- Building Organization Culture and Values – An exploration of how teams and organizations define their culture and the critical values that are the foundation of this culture.
- Executing Strategies and Tactics – An exploration of how teams and organizations develop and execute strategic and tactical plans to achieve the purpose, goals, and objectives of the organization.
- Utilizing Team Norms and Social Contracts – An exploration of team dynamics and the use of norms and social contracts to drive team performance.
- Attracting, Hiring, and Retaining Diverse Talent – An exploration of how we ensure that the right people with the right skills are in the right positions on teams and within organizations.
- Creating Agile Teams and Organizations – An exploration of how we lead change and develop agile teams and organizations that are able to morph to remain effective over time.

These elements are the foundation of leading teams and organizations to success.

Finally, Part 6 of "US," **Leading an Inclusive Construction Industry,** explores how each of us is responsible for continuing to shape the future of our industry. Included are specific suggestions for how to lead the industry, including strategies for joining others in affecting systemic change.

Throughout the book, I focus on describing the inclusive industry I believe we all desire.

After the first day of an Australian Pipelines and Gas Association (APGA) women's leadership program, all 26 women boarded a bus to attend an APGA networking event at the Adelaide Zoo. This was our first experiment with taking an entire women's leadership group to an APGA networking event, and

> I was interested in seeing how it would turn out. I would guess that there were about 30 women and 30 men at the event. Had we not brought the women's leadership group, it would have looked much more like a typical event, with five women and 30 men in attendance. I was a bit worried it was going to resemble a 1950s high school dance with women on one side, and men on the other. But to my relief and joy, it looked like a normal event in the world, with groups of men and women talking naturally to one another. It was a heartening sight.

This is inclusion—"the act or practice of including and accommodating people who have historically been excluded (because of their race, gender, sexuality, or ability)" ('Inclusion,' 2024). Inclusion is an action, and it is deliberate. It is thinking, "Perhaps this person does not feel like they belong. What can I do to help?" Building an inclusive team, organization, and industry starts by valuing diverse people and the unique perspectives each brings to our thinking and problem solving. To be an inclusive leader, you talk about diversity and you measure diversity. You also ask people about their experience on your team and within your organization. You create a safe space for people to share whether they feel like they belong or not, and what actions their teammates might take to help them feel like they belong. You also have zero tolerance for behavior that challenges inclusion. Period.

> As author and social justice advocate Howard Ross shares, "*Much of the work that was being done in diversity was events. We'll do a diversity training. We'll have a black history month celebration and an international foods day celebration in a cafeteria, and we could check off a diversity program. We were approaching this in a way that we wouldn't approach any other serious topic. Most construction companies that I've interacted with have safety protocols and those safety protocols are in every aspect of the organization. We buy products that are safer, we train people on how to use them safely, we have reporting mechanisms, we know what to do when somebody breaks the rules and puts other people in danger. There are hundreds of ways that we build a safety perspective for organizations that do it well and keep people safe. It's not just, 'do these quick three things and check it off.' People need to take the same kind of systemic look at our organizations relative to diversity and inclusion and accessibility.*"
>
> (Howard Ross, speaking to Gagel, 2023 Greatness Podcast)

As you read this book, I encourage you to deeply reflect about yourself as a leader, including your natural gifts and challenges, your aspirations as a leader, and how your leadership can help us build a more inclusive industry. One of my PhD professors advised us that some books are to be devoured and some are to be nibbled. I encourage you to take one of these two approaches for this book: devour it in its entirety and then go back and nibble on sections you would like to further explore, or nibble on the book topic by topic, taking time to reflect upon each topic that seems important to you at this point in your leadership journey. You might document a leadership "playbook" of the ideas that are important to you. I also hope this book will inspire you to dive deeper into one or more of the topics I have included via other resources.

To learn and grow, you must act upon what you learn. This is not a rhetorical exercise; this is about becoming a better leader in leading yourself, your team, your organization, and the industry. Leadership is not about a title; leadership is about the ability to influence others and help others succeed. It requires practice and trying new things. I suggest you engage with a mentor, sponsor, coach, or accountability buddy of some type as you hone your leadership skills. Share your reflections from the book with that person and tell them what new tactics you are trying as a leader, as well as what skills you hope to enhance. Speak with them about what has worked, what has not, what adjustments you will make, along with what new things you will try.

I have many favorite books that I periodically revisit when I need to be reminded of the wonderful ideas shared in them. I hope this book is that for you—a book you keep on your shelf and sometimes think, "I need a refresher on this topic." As I say to my clients and students, I am always here for you… and I see a great leader in all of you!

References

Gagel, G. and Noonan, A. (2022). *Andre Noonan discusses gender diversity in construction* [Podcast]. 7 January. Available at: https://open.spotify.com/episode/5aM6gnz96oIbulicBcEI5Z [Accessed: 1 May 2024].

Gagel, G. and Ross, H. (2023). *Howard Ross discusses unconscious bias and inclusion* [Podcast]. 30 June. Available at: https://open.spotify.com/episode/626OFIWKg1RBJr8CWFC6iq [Accessed: 2 May 2024].

'Inclusion.' (2024). *Merriam-Webster Online Dictionary*. Available at: https://www.merriam-webster.com/dictionary/inclusion [Accessed: 1 July 2024].

LaRocco, L. A. and Goldberg, N. R. (2023). '*The iconic American hard hat job that has the highest level of open positions ever recorded*,' [CNBC]. 29 July. Available at: https://www.cnbc.com/2023/07/29/the-hard-hat-job-with-highest-level-of-open-positions-ever-recorded.html [Accessed: 1 July 2024].

McDowall, D. (2023). Interview by Gretchen Gagel [Zoom], 30 November.

'What makes a high-performing team? The answer may surprise you.' (2017). *[MIT Management Executive Education]*, 22 March. Available at: https://exec.mit.edu/s/blog-post/what-makes-a-high-performing-team-the-answer-may-surprise-you-MCIE4TDCCNFZAL7MK6DYC2S6TEK4 [Accessed: 1 July 2024].

Part I

"Me" – The Lived Experience of the Construction Industry

1

Exploring Unconscious Bias and Dominant Culture

Introduction

Recently, I took the ferry to Raymond Island in Australia to see the koalas that were introduced to the island in 1953. I had visited the island twice before, and knew that over 200 koalas lived there. Easy to see, right? Not that afternoon. After about 25 minutes of walking on the "koala trail," I thought to myself, "I'm never going to see one!" Then I remembered the advice of one of my PhD professors, Dr. Susan Lynham, originally from South Africa: you must have "soft eyes" to find the animals. I relaxed and focused on my "soft eyes," and yes, there was one koala—and another, and another.

I believe concepts such as unconscious bias and dominant culture in the construction industry require those same "soft eyes" to see, and that understanding these concepts is a critical step in creating a more inclusive industry. In the US, men make up approximately 90% of the construction industry, and many of us grew up in a patriarchal society (Gallagher, 2022). These facts contribute to the unconscious bias and dominant culture of the construction industry that causes some people to feel as though they do not belong. This is not a slam of men. Most people are not mean-spirited and are not deliberately trying to make people feel like they do not belong. Men, if you are reading this book, you are amazing because you are seeking to learn more about the lived experience of those who do not look like you, and you want to understand how you can help women thrive in the construction industry. Men may feel like parts of the world and our industry are against you, and I hope you realize we are not.

> "My advice would be to embrace change in our industry. Sometimes, unfortunately, human nature being what it is, your first reaction can be protectionist of yourself and you're looking inward. Don't fight what we're talking about here, embrace it,

> *be a champion of it, accept it and don't feel as though, 'oh well, it's fifty percent of the opportunities' because that's just a pure evolutionary thing. Fundamentally it makes good business sense."*
>
> (Andre Noonan, speaking to Gagel, 2022 Greatness Podcast)

In the book *Immunity to Change*, authors Kegan and Lahey refer to three stages of adult learning that I feel are important to consider in this discussion of unconscious bias and dominant culture (Kegan and Lahey, 2009):

1) **The Socialized Mind** – About 58% of adults operate with a socialized mind, meaning their actions are dictated by the social norms of the groups they operate in. If you spend most of your time with people who hate, you may believe it is okay to hate. You are a team player and follower who looks to others for direction and social norms.
2) **The Self-Authoring Mind** – About 35% of adults operate at the self-authoring mind level, meaning their actions are determined by the communities they operate within and the decisions they reach through their own critical thinking. You have your own perspective on the world and how you want to live in it, and you probably influence the thinking of others.
3) **The Self-Transforming Mind** – Only 1–2% of adults reach the self-transforming level of thinking. I describe this level as almost being able to step outside of your body, to view yourself in the world and deeply reflect upon your actions, your decisions, your feelings, and the impact you have on others. I believe that many people think they are operating at this level when they are in fact operating at one of the first two levels of thinking.

I believe that the pace of today's world and the amount of information we are bombarded with each day create an even greater barrier to operating with a self-transforming mind because this level requires time spent on reflection, and that time is difficult to find. Having said that, striving as leaders to think critically and be present to the lived experience of others is important.

The construction industry of today is very different from what I experienced 40 years ago. We are not an industry that is "broken." We are an industry that is evolving—that is becoming more inclusive because we understand that diverse thinking is better thinking—and that is exciting. There is room for improvement, and a deeper understanding of people's lived experiences (including unconscious bias and dominant culture) will help us all.

As James Breen, Vice President Pharmaceutical Global Engineering and Technology for Johnson & Johnson, shares: "*We have a long way to go in our industry,*

> *but we've made huge progress since the '80's. People are not clear on what inclusion is, that it is much more than gender. Women are half the population and have half the good ideas, and we need to leverage that as an industry."*
>
> (Breen, 2023)

As you read through this chapter, I encourage you to ask yourself these questions: "Do I follow the lead of others, or do I think for myself? Am I presenting nuanced concepts like unconscious bias as they unfold before me, or do I walk by without noticing? Am I present to the lived experience of people who do not look like me?"

The Lived Experience of Women in the Construction Industry

Before we dive into the topics of unconscious bias and dominant culture, it is important to understand how these concepts shape the lived experience of women in the construction industry. While this book is intended as a positive, uplifting contribution to your leadership journey and an exploration of how we all help women and other underrepresented communities thrive in the construction industry, I would be remiss if I did not explore the challenges of women in our industry.

> I found this comment by a respected woman leader in our industry during our interview for this book shocking: *"Aggressive is worse than crude. I will sit through locker room talk, but when people raise their voice, curse at me, that is where I draw the line. The last time I was physically threatened at work was in 2001 when he came over the table. Now it's happening again the last couple of years, post Covid with more stress in the environment. I'm being cursed at by clients, physically threatened. Maybe it's just the projects I'm on right now."*
>
> (Anonymous)

During my first two years as a manufacturing operations manager in the 1980s, I had multiple experiences of sexual harassment and bullying. During arbitrated union negotiations, I walked into the processing portion of the plant to find "F--k Gretchen" painted in six-foot lettering on a large processing tank. I politely and firmly asked one of the processing operators to call maintenance to have it painted over. One of the supervisors who worked for me walked up behind me while I was sitting at my desk and kissed me on the back of the neck. I politely and firmly asked him to never touch me again. A maintenance worker I had refused to date started yelling at me in the plant one day. I politely and firmly asked him to meet with me with his supervisor. During that meeting, he told me that I had walked by him without saying "hello," that I acted better than him, and that God said we were all created equal. I politely and firmly stated to him, in front of his supervisor, that I was not ever going out with him, that within this plant we were not equals, and that it was not okay to treat me disrespectfully.

When I was transferred to manage a cereal plant in Davenport, Iowa, in 1988, my first shift supervisor insisted upon calling me "boss lady" and dropped as many f-bombs as possible until he realized it was okay to have a woman as his manager. During the plant's nine-hole Tuesday night golf league, I was constantly asked if I was going to join the men for their weekly visit to the strip club afterward. One day, I finally said, "Great, yes, I'll be happy to join you on the last night of the golf league." That final evening during golf, they joked around with each other, saying, "Hey, Gretchen's coming tonight." The looks on their faces when I calmly walked into the strip club were priceless. I had one drink and left. I had made my point—don't ask for it if you don't want it.

As I transitioned into management consulting in the construction industry, I experienced more of the same—late-night calls to my hotel room from men attending my meetings, and a construction industry executive placing his hand on my thigh, twice, during a dinner in Houston. After lifting his hand off the second time, I politely said to him that if he wanted a fork in the back of his hand, he would do it one more time. Sadly, I felt the need as a 50-year-old woman to ask one of the other men at the dinner to walk the two blocks to our group's hotel with me, as I did not feel safe. One day, one of our administrative assistants overheard a fellow consultant speaking to a potential client about me—"Don't worry, she's not too good-looking, she just had two kids and is a bit of a battle-ax." He was reminded by our leader of the inappropriateness of his words.

Those are the egregious examples. The more subtle behaviors were at times more challenging, such as people assuming I was an administrative assistant instead of a consultant, or questioning if I had the knowledge required to help them just because I was a woman.

What gave me the fortitude to stay in the construction industry? I believe my father telling me I could not attend engineering school laid the groundwork. No one was going to tell me what I could or could not do; no one was going to ruin my career because of their bad behavior. I just would not let it happen. I also love the construction industry. I love the problems we solve and the amazing contribution we make to society. And so, I stayed.

I also realize that my lived experience in the construction industry is different from the lived experience of women of color in the construction industry. Friend and construction industry leader Denise Burgess has recounted her frustration with people assuming her father had sent her to take the meeting notes, and of having to constantly reprove herself as a Black woman. *"I was doing a project in Utah, and I walked into a meeting of all men, and they looked at me and said, 'Oh, we thought it was Dennis, we thought it was a typo.' I was negotiating a contract and they said, 'Did you want to wait for someone?' I'd think, did they not like me because of my gender or not like me because of my race? And my dad would say, does it matter?"* (Denise Burgess, speaking to Gagel, 2023 Greatness Podcast).

Friend and construction industry leader Suzanne Arkle, a woman of color who is African American/Japanese, shares this about her experience as a woman in construction: *"We were looking at the metrics of participation for this particular owner who wanted to look at race, and I said something about being an African American businesswoman. She looked at me and she said, 'You're African American? I thought you were Oriental. That's why you're so smart!' I was taken aback but I knew it was not coming from a bad place. The first thing I said, "It's not Oriental anymore … you can say Oriental rug but not Oriental person.' She was embarrassed but I knew that she would receive it"* (Suzanne Arkle, speaking to Gagel, 2024 Greatness Podcast).

I met Laura Miranda, co-founder of the You Don't Look Like an Engineer Podcast, during a construction conference where I heard her story of coming to Australia as a young woman without strong English skills. *"Just be authentic as to who you are. Sometimes your career path doesn't make sense to other people, and it doesn't have to as long as you're happy with the decisions you're making, and you are being authentic to yourself. They don't have to behave like the boys to fit in and I think when we talk about leadership that's one of the things that sets us apart from men"* (Laura Miranda, speaking to Gagel, 2024 Greatness Podcast).

> Construction User Roundtable (CURT) Chair and Ellisian CEO Jim Ellis put it well during our interview for this book: *"Women, African Americans – we just flat out don't pay attention to it. Why would they want to work in the good old boy's club, deal with graffiti, not having the right facilities? Look at the numbers."*
>
> (Ellis, 2023)

These are the experiences of women in our industry that we need to be present to. These are the actions, the words, and the behaviors that make us feel as though we do not belong. In most instances, it is not meant to intimidate or make us feel uncomfortable or unwanted, but it does.

> **Activity:**
>
> - Take time to reflect upon your lived experience in the construction industry. What experiences come to mind most quickly? How do these experiences shape your thinking about the culture of our industry?

- Take time to reflect upon the people who do not look like you in the construction industry. What is their lived experience in the construction industry?
- Invite a person who doesn't look like you out for coffee. Let them know that this book has prompted you to be more curious about the lived experience of people who are not like you. Ask them gracious questions. "What is it like to be a (woman, man, person of color, immigrant) in the industry? What do others do that makes you feel like you belong? What do others do that perhaps make you feel like you don't belong?"

Understanding Unconscious Bias

For years, I have been going out to dinner with men—clients, partners, and colleagues—and in the United States, it has been customary to hand your credit card to the waitstaff, who then return it with the bill to be signed. I'm going to estimate that 95% of the time when I hand my credit card to the waitstaff, they return the bill and card to one of the men sitting at the table. The card says "Gretchen" on it! I was sharing this story with the male CEO of a construction firm, and he joked that when he goes out to eat with his wife, they always give him the steak and her the salad, even though the order is reversed. Recently, a shared ride driver asked me about my week, and I recounted that I was traveling from Kansas City to Chicago to DC to New York City and back. He immediately asked me if I was a flight attendant. These are everyday examples of unconscious bias.

Bias is defined as "**1a:** an inclination of temperament or outlook; *especially*: a personal and sometimes unreasoned judgment: **PREJUDICE; b:** an instance of such prejudice" ('Bias,' 2024). We all have bias, and we all need bias to help us sort through the 34 gigabytes of data and information—or about 100,000 words—we are bombarded with every day (Wonder, 2022). Our brain is efficient in sorting through this data in part because our biases help us decide what information is important or not important, and what is going to kill us or keep us safe. Most of this happens without us thinking about it, and this creates efficiency. I think of bias like file folders in our brain—the systems and mental models that help us make sense of our chaotic world.

"We do hundreds and hundreds of things every day that we don't need to think about. Daniel Kahneman, the brilliant psychologist that won the Nobel Prize for

> *the work that he and Amos Tversky did in understanding the brain and how it works, uses a great metaphor for it. You're driving to work in the morning. The brake lights come on the car in front of you and your foot hits the brake pedal before you even have a chance to think about it because over the time of driving and learning to drive you've learned that brake light means danger and you could crash into that car. You wouldn't go to the slow, more thoughtful part of your brain and say, 'What am I supposed to do in this circumstance?' because if you did, you'd be in their back seat right? Our brains have developed over time to make these kinds of quick decisions to help us move through life more quickly, to not have to remake decisions on a constant basis and to deal with everything as if we've never seen it before and to keep us safe."*
>
> (Howard Ross, speaking to Gagel, 2023 Greatness Podcast)

The term "unconscious bias," first used in 1995 by Mazarin Banaji and Anthony Greenwald and often referred to as "implicit bias," refers to the underlying assumptions we have—typically about a specific group of people—which impact our behaviors and decisions (Greenwald and Banaji, 1995). Often thought of as "stereotyping," unconscious bias is unintended and different from explicit prejudice.

> *"The second piece was a realization that came from a lot of the research I've been doing on how much of the bias that we attribute to people's intention was not in fact intentional at all, that in fact overwhelmingly unconscious bias is what drives people's behavior, and how important that is because it fundamentally changes the nature of how we interact with people. If I believe that you're doing something to me with intention, my way of interacting with you is very, very different than if I believe that you may not realize you're doing that, or you may have a blind spot about that."*
>
> (Howard Ross, speaking to Gagel, 2023 Greatness Podcast)

Understanding unconscious bias and its impact on our behaviors and decisions is critical in helping us build a more inclusive industry because as leaders, we must be aware of the assumptions we make about people and groups of people. Understanding these assumptions and pausing to think about them provides us with an opportunity to change our behavior. For example, the waitstaff could pause and think, "Am I handing this credit card to the man at the table out of habit? Because of an assumption? Maybe I should pause and look at the name on the card, or pause to remember who handed it to me before automatically handing it to the man at the table."

Harvard professor and psychological safety expert Amy Edmondson shares these thoughts: "*Another interpersonal skill is recognizing and being aware of cognitive biases and finding ways to stop and pause and reflect and reframe because we know we might be missing something, or we know we'll be misinterpreting something the first time around.*"

(Amy Edmondson, speaking to Gagel, 2019 Greatness Podcast)

I am not referring to explicit prejudice, harassment, or bullying. Explicit prejudice is defined as "a negative attitude against a specific social group that is consciously held, even if not expressed publicly" ('Explicit Prejudice,' 2018); harassment is defined as "the act of making unwelcome intrusions upon another" ('Harassment,' 2024); and bullying is defined as "The repetitive, intentional hurting of one person or group by another person or group, where the relationship involves an imbalance of power. Bullying can be physical, verbal, or psychological. It can happen face-to-face or online" (Anti-Bullying Alliance, n.d.). If we see explicit prejudice, harassment, or bullying, we need to call out that behavior with zero tolerance. We must, as leaders in our industry, stamp out hate and prejudice. Allowing any of these behaviors violates respect for human beings—a fundamental condition of the civilized world.

As I described earlier, women in the construction industry frequently face unconscious bias. When I first started with FMI and traveled with my male counterparts, it was often assumed that I was an administrative assistant. Even later, when I was CEO/President of Continuum Advisory Group and traveled with a male associate, it was assumed that they were the more senior person in the company. I was often asked if we were a woman-owned business, and when I responded "no," the next question was frequently, "Hm, so why did they hire you to be CEO/President?" I would politely respond that they had hired me as President because I was the most qualified person to be CEO/President of the company.

"*I have seen that for women, particularly contractors, if you're not able to sort of step up and step out and state your case you will get walked over because there will also be an assumption that you don't know as much; and even in today's world I see it. I met with a woman contractor this morning, a supplier, fourth generation, the company started in 1900, she is the fourth generation. She said, 'If I could just change the perception a little bit, shift it a little bit, that I know what I'm talking about, that would be so helpful to me'.*"

(Suzanne Arkle, speaking to Gagel, 2024 Greatness Podcast)

Kim Neuscheler (2023), Vice President and General Manager with Turner Construction Company, shared this story: "*Anthony, an elevator operator on one of our jobs, knew me as Kim but didn't know I was running this billion-dollar project. We were on our last big push, and I ordered breakfast for the team. The elevator operator asked a question of the Superintendent who indicated, 'you need to ask the big boss that question'. Anthony said, 'Why didn't you tell me you were the boss?' I said, 'I didn't need to, you always treated me with respect.' But he assumed I wasn't the boss.*" Kerri Smith Petrillo (2023), Chief Talent and Strategy Officer at Baker Construction, said it well: "*Did I have to work harder than my male counterparts to achieve the same success? Probably, but it makes it that much better!*"

Unconscious bias is important to understand because it creates assumptions that hinder our ability to be inclusive. "*Let's say I've got this position open for a leader in the construction industry and I really do want to have more women in leadership. You've got this great resume but, in the interview, you're not responding like the guys. Maybe you're a little bit quieter, you don't use as many curse words, you're a little bit more formal, you don't have that executive presence thing that I'm expecting to see because unconsciously I'm comparing you to all the male leaders I've seen. Somebody says to me, 'How did it go?' and I say, 'Well, she looked great on paper, but she just didn't have that executive presence,' and I don't even realize that it's my expecting you to act like a man that is having me show up that way. You go to the company next door and become their top executive*" (Howard Ross, speaking to Gagel, 2023 Greatness Podcast).

> Unconscious bias is closely related to another important concept, privilege, as shared by Lee Jourdan, Director and retired Chief Diversity Officer for Chevron: "*As an example I can talk about my own privilege. I'm an able-bodied, cis-gender, Black man. Being able-bodied I don't have to approach a building and be concerned about access to those facilities. I don't have to worry about my workstation being configured for me, it's something that never enters my mind. I am heterosexual. I could drop my kids off at school and never have to worry about them being bullied because they had two moms or two dads. Being black, the privilege I see there, particularly in the last year or so where people talk about Covid … , people would use 'Chinese flu' for example, and this created a lot of hate of the Asian community and a lot of my colleagues who are Asian were concerned about walking in the park or going to the grocery store. I didn't have to worry about that and so again it was a privilege that I had. Lastly, being a man, that created an opportunity for me to get the benefit of the doubt when women did not. Being a Black man, I would get the benefit of the doubt.*"
>
> (Lee Jourdan, speaking to Gagel, 2021 Greatness Podcast)

The good news is that we are starting to talk about unconscious bias a great deal more in the construction industry. Back in 2022, I was having some fun with the "demolition" theme of the magazine *Inside Construction* during my regular editorial and wrote a piece called *Demolishing Unconscious Bias in the Construction Industry* (Inside Construction, n.d.). Amazingly, the Australian Institute of Quantity Surveyors called me after reading it and asked if I would do a webinar on the topic of unconscious bias for their membership. It was a happy moment for me, as I realized that more people in our industry wanted to understand unconscious bias.

I believe these are the important steps to being present to unconscious bias:

1) **Learn About It** – Learn what unconscious bias is and how to be present to the unconscious bias we all have; and encourage those around you to learn about it as well—your team, your peers, and your organization.

2) **Set a Goal** – Set goals personally, as a team, and as an organization to talk about unconscious bias and its experiences. There are many tools available to help lead these discussions.

3) **Be Open to Feedback** – Be open to (and solicit feedback from) those who may be able to safely inform you about times when unconscious bias could be influencing your decisions or behaviors.

4) **Create Psychological Safety** – When people feel psychologically safe, they will point out examples of unconscious bias. I politely asked the Uber driver who had just asked if I was a flight attendant, "Would you have asked me that question if I were a man?" and he admitted that no, he would not have asked me that question had I been a man. I had politely helped him understand his unconscious bias.

> Bob Nussmeier, Vice President of Strategic Clients, Baker Construction, shares: *"Men should embrace servant leadership and support the success of women in our industry. We need to address underlying bias and I'm glad the chauvinistic field behavior is declining."*
>
> (Nussmeier, 2023)

It requires courage to be the person who raises your hand to point out unconscious bias. As industry leader Dr. Barbara Jackson says, *"My advice to men in our industry is to be conscious, to intervene, to ensure that women's ideas are heard and acknowledged. If the woman is being interrupted say, 'I didn't catch that,' intervene in the discounting of women, sound the alarm about unhealthy conditions"* (Jackson, 2023). Ornella Houssais (2023), Assistant Project Manager with Brinkman Construction, offered this advice: *"I would encourage men in our industry to imagine that it is their daughter in the room and think about how they would want them to be treated. It helps them become more aware."* Small actions matter as well. As industry leader Lisa Hinz shares, *"Please do not talk over women, truly listen to what they are saying. When you dismiss women's ideas you are feeding unconscious bias. Small and simple things can make a big difference"* (Hinz, 2023).

> Sometimes, overcoming unconscious bias starts with intentionality, as was exhibited by the construction project executives for the recently constructed KC Current women's soccer team stadium in Kansas City. As Erica Jones, Vice President of Marketing for construction company Cerris (formerly MMC Corp), states: *"At our last On The Rise program we did a panel about the KC Current Stadium and how intentional it was for those owners to make sure that their entire team that was putting together the KC Current Stadium was female-led, female engineers, architects, construction, as many as possible, to bring in as many minorities as possible too. I think that without that intention we aren't going to move forward in our industry."*
>
> (Erica Jones, speaking to Gagel, 2024 Greatness Podcast)

Everyone has unconscious bias. We need bias to sort through information, and we need to be present to unconscious bias to help us understand the behaviors we are not aware of that cause others to feel they do not belong in the construction industry. I encourage you to continue learning about unconscious bias and to help those around you in the construction industry learn about it as well.

> **Activity:**
>
> - Think about instances where you have made assumptions about people or groups of people. What was driving those assumptions? What unconscious bias do you have about certain groups of people?
> - What might you do to learn more about and be more present to your unconscious bias? How might you help your team or organization learn more about unconscious bias?

Understanding Dominant Culture

Several years ago, my daughter, her boyfriend, my sister, and I took an overnight flight from Dubai to Kigali, Rwanda. From the outset, it was clear to me that we had no idea what we were doing. Everyone checking in for the flight looked different from us and knew exactly what to do—where to take their bags and how to have their bags and boxes weighed. Our boarding passes indicated that boarding commenced two hours before the flight, which I assumed had to be a mistake.

When we walked up to the gate 90 minutes before the flight, we were the last in line to board. Every piece of hand luggage was being weighed, and my backpack almost did not make it onto the flight. The cabin looked like a cargo plane, with boxes in every open seat. I felt completely out of place and uncomfortable during this entire experience, like I was in some type of movie where I did not belong and did not know how to accomplish tasks familiar to me—especially given my frequent travel.

The term "dominant culture" describes the important values, predominant characteristics, belief systems, and behaviors of a specific group of people ('Dominant Culture,' n.d.). During this flight, I was experiencing a dominant culture that was unfamiliar to me and that I was not a member of. Many of us go through our entire life without this experience, and that makes it difficult for us to understand and have empathy for people who are not part of the dominant culture and who perhaps do not feel they belong. Dominant culture is at times invisible to us when we are within it.

When we think of the dominant culture of the construction industry, in most parts of the world, we think "white male" and "patriarchy." Again, this is not a slam of white males, and it is how our industry has evolved because of the population in the construction industry.

> *"This is a business designed by men for men and the culture is killing people off. The younger men, 35–45, they are overworked, and they are leaving. We treat people like a commodity. It's not that the leaders don't care. They don't know what to do so they don't do anything, they lack vision and creativity, vulnerability, all the stuff that's needed to shift to a more humanized culture."*
>
> (Jackson, 2023)

I have tremendous respect for the work of Bill Proudman and Dr. Michael Welp, co-founders of White Men as Full Diversity Partners (WMFDP), who have been helping white men understand dominant culture for 30 years (Welp, 2016). Sam Clark, CEO of Clark Construction in Lansing, Michigan, and third-generation general contractor and Chair of the AGC of the American Diversity and Inclusion Committee, gave a compelling presentation to this committee about his work with WMFDP. Sam explained how he attended their White Man's Caucus training and realized for the first time what it felt like for those people **outside** of the dominant culture. Sam then understood why their company's DEI efforts had failed and sent many of their leaders through training. The associated business results were compelling, including growth from $200 million to $500 million in revenue in three years, a 99-employee engagement score, and women literally beating down their door to go to work with them (Sam Clark, speaking to Gagel, 2023 Greatness Podcast).

Sam Clark, CEO of Clark Construction, Discusses the Cultural Transformation of His Company

Sam Clark began our discussion of the cultural transformation of his organization by discussing a conversation he had with one of their female employees: *"I went to talk to her about it one day, just some issues that I saw in a survey, words like 'group think' and 'bro culture', and that sort of thing, and I had no idea really what that was. She said, 'There's issues here but they're not issues that are any different than any other company and this is a good place to work.' I think she just didn't have the bandwidth to try to explain it to me because I wasn't going to get it and that didn't sit well with me. After going through that program and learning about dominant culture and how we can change that and be more inclusive, it really dawned on me that trying to introduce diversity into a company that doesn't have an inclusive culture was backwards and it just wasn't going to work. We had a theme for the year of 'inclusion,' and we talked about that at every meeting and did some training on it; and it's like anything, it feels like you're not moving the needle very much and then all of a sudden things start changing in big ways.*

The one light bulb was the realization that some of the things that we do cut into the dignity and productivity of other people, and particularly women. We have a lot of women here, and I thought, we're putting all these hurdles in front of women in small ways and some big ways. Every day these little things that distract them from their work, when people are making comments about their appearance or when they're talking over them, not listening, not recommending them for promotions, maybe in their reviews mis-assessing their talents because women tend to be more reflective and better risk managers, and that can show up as 'you're not willing to take a chance', not really recognizing the value in the way women think differently. And that's tiring. So if we can make our leaders more aware and chip away at some of that ... we can't fix all the world's problems that are affecting women and we have to recognize that I can't affect how other people in the world treat women, I can at the company. And so, my thought process was if we can knock down some of these hurdles, women can be a lot more effective and we will attract all these talented women in our business, because I don't know very many construction companies that are paying attention to this.

That's a key. It's really not white male bashing. When we talk about white male culture or dominant culture, it's not good or bad. All those traits are positive but if you utilize your strengths all the time they become a weakness. It's helping white men also recognize the box that we put ourselves in or that we're put in by the dominant culture. Not everybody fits neatly into the dominant culture including white guys. It's allowed a lot of men in our company to be themselves as well. I brought the leaders in our company together and helped them open up and learn

> *things about one another and learn about the culture and why we're doing this, and it really changed the lives of a lot of people.*
>
> *But early on after I came back with this information and was trying to get this ball rolling, there were a lot of people understandably that didn't understand why we were doing this. I had an executive come in my office on a Friday afternoon and we spent about two or three hours talking about it because he's getting questions from people like, we have all these problems, why are we focusing on diversity? That's not even a problem in their minds. I did my best to explain it to him, and admittedly he said after he went to the caucus, 'I didn't really understand your vision. I thought I did but I was only getting about five or ten percent of it. This stuff is amazing and it's critical to the businesses.'"*
>
> (Sam Clark, speaking to Gagel, 2023 Greatness Podcast)

This work is about understanding dominant culture as it is experienced by those **inside** of the dominant culture and by those **outside** of the dominant culture. When you are in the dominant culture, you are "in the bubble." You may not understand that you are in the dominant culture because you are **in** the dominant culture. You also do not fully understand the lived experience of those people **outside** the bubble. Being outside the bubble looks like the guys going off for drinks and not inviting you, maybe talking about a big deal and you are not involved in the conversation. It looks like men joking around until you walk in the room, then they suddenly stop and look at anything but you. It looks like a man turning to you as the only woman in the meeting and asking you to take the notes, and frowning when you suggest one of the men do it. That is how I felt for many years in the construction industry, as an outsider trying to fit in. When I joined FMI, I wore khaki pants, blue blazers, and penny loafers just like the guys. I wore my hair short and tried not to look too feminine. This is not just an issue for women. I recently spoke with a talented, brown-skinned male who was recruited into our industry. After three years, he is considering leaving because he does not see a commitment to inclusion at his organization and does not believe he fits in.

Being present to the dominant culture in an industry, an organization, or even a project team helps you understand what it feels like both for those who resonate with the dominant culture and those who may not. Those who more closely align with the dominant culture might feel as though they are being invaded or blamed for the dominance of that culture. Those who are outside the dominant culture will probably feel left out and as though they do not belong. If this is pervasive, these people will likely leave to seek out a place where they feel like they belong.

Creating an inclusive culture within your team or organization recognizes and acknowledges the dominant culture, and then works to shift the overarching culture of the team or organization such that everyone feels they belong.

The important takeaway from this is that every leader in the construction industry should take the time to reflect upon the dominant culture of their team, their department, their organization, and our industry. Only through this awareness will we all be present to what we as leaders do to either reinforce the dominant culture and drive those who do not fit into it away or break down the barriers of the dominant culture by being intentional in building an inclusive industry. We are making progress, as Tom Reilly, President, Turner Construction Company, shares: "*We have more awareness as an industry. Turner was a founder of 'Inclusion Week' in the construction industry, and it is an important concept for us. The first year 1,000 companies participated, this year 5,000 companies participated. Awareness and momentum for change is building*" (Reilly, 2023).

Activity:

- Take a moment to reflect upon the words you would use to describe the dominant culture of your team, department, or organization.
- Who do you think feels they belong in that dominant culture? Who does not feel they belong?
- What steps might you take as a leader to help those around you understand the importance of understanding dominant culture and its impact on the people on your team?
- What might you do as a leader to be more intentional about including people who may not feel they belong in that culture?

Conclusion

Lee Jourdan shared the story of his experience with a male executive colleague that I believe exemplifies the impact of unconscious bias and dominant culture on our ability to build a more inclusive industry: "*I was an executive at a Fortune 10 company and was talking with another very senior executive, and he kind of leaned in and he said, 'You know, Lee, I can tell you this', and he felt he could tell me this because I was a man. He said, 'There are just some jobs that women can't do.' I thought, wow, maybe he's talking about some physical requirements, lifting a heavy pipe or something like that, which still didn't make sense because I know a lot of women that are stronger than men. He said, 'Yeah, the business that I'm in, I may have to call someone on a Sunday night at 9:00 PM and tell them they have to be on a plane Monday morning at eight o'clock, and women just can't do that.' I'm thinking, 'Why?*

That you think they need to make sandwiches or something? I mean, why?' and I said to him, 'Well, why can't you just at the outset make clear what the expectations of the job are and if she says she can do it, then she can do it just like a man would do' and he just kind of waved me off. I think to myself, this is a very senior individual who is well into his sixties who has gone through his entire career thinking that way, making assumptions about what women can't do" (Lee Jourdan, speaking to Gagel, 2021 Greatness Podcast).

> *"My advice to men in our industry – don't make assumptions based upon what you think she is capable of doing, or what you think she is interested in. It would be really great if they asked."*
>
> (Neuscheler, 2023)

This is not just yesterday's problem; this is today's problem. I envision an inclusive industry that reflects the make-up of the communities we serve and is grounded in a dominant culture of respect that brings about a sense of belonging for everyone. This vision is achievable, and each of us taking the time to learn more about dominant culture and unconscious bias will provide the knowledge we need to help make that vision a reality.

References

Anti-Bullying Alliance. (n.d.). *Our definition of bullying*. Available at: https://anti-bullyingalliance.org.uk/tools-information/all-about-bullying/understanding-bullying/definition [Accessed: 20 May 2024].

'Bias.' (2024). *Merriam-Webster Online Dictionary*. Available at: https://www.merriam-webster.com/dictionary/bias [Accessed: 20 May 2024].

Breen, J. (2023). Interview by Gretchen Gagel [Zoom], 1 November.

'Dominant Culture.' (n.d.). *Cambridge Dictionary*. Available at: https://dictionary.cambridge.org/example/english/dominant-culture [Accessed: 20 May 2024].

Ellis, J. (2023). Interview by Gretchen Gagel [Zoom], 25 October.

'Explicit Prejudice.' (2018). *APA Dictionary of Psychology*. Available at: https://dictionary.apa.org/explicit-prejudice [Accessed: 20 May 2024].

Gagel, G. and Arkle, S. (2024). *Suzanne Arkle discusses DEI in the construction industry* [Podcast]. 22 March. Available at: https://open.spotify.com/episode/0xZl58QBnViz68wm0hcj4Y [Accessed: 2 May 2024].

Gagel, G. and Burgess, D. (2023). *Denise Burgess discusses her career as an African American woman leader in the construction industry* [Podcast]. 29 December.

Available at: https://open.spotify.com/episode/77UCAi1FXUKeMOS6yfes8z [Accessed: 2 May 2024].

Gagel, G. and Clark, S. (2023). *Sam Clark discusses diversity equity and inclusion in the construction industry* [Podcast]. 21 April. Available at: https://open.spotify.com/episode/6pSdP91DlpLsBVX6OmtokB [Accessed: 2 May 2024].

Gagel, G. and Edmondson, A. (2019). *Amy Edmondson discusses her book The Fearless Organization on psychological safety* [Podcast]. 30 October. Available at: https://open.spotify.com/episode/6joytIVhV3XlGFZx6rcodC [Accessed: 28 April 2024].

Gagel, G. and Jones, E. (2024). *Erica Jones discusses women's leadership in construction* [Podcast]. 23 February. Available at: https://open.spotify.com/episode/3RQth0SyIU9nWVsm4RSqKg [Accessed: 2 May 2024].

Gagel, G. and Jourdan, L. (2021). *Lee Jourdan, former chief diversity officer at Chevron, discusses privilege, diversity and inclusion* [Podcast]. 29 October. Available at: https://open.spotify.com/episode/05qjBlrUC0tjWXkUoVZlpk [Accessed: 1 May 2024].

Gagel, G. and Miranda, L. (2024). *Laura Miranda discusses her podcast "You Don't Look Like an Engineer"* [Podcast]. 8 March. Available at: https://open.spotify.com/episode/6sLfjJDdlYzWoT5zkSD1nR [Accessed: 2 May 2024].

Gagel, G. and Noonan, A. (2022). *Andre Noonan discusses gender diversity in construction* [Podcast]. 7 January. Available at: https://open.spotify.com/episode/5aM6gnz96oIbulicBcEI5Z [Accessed: 1 May 2024].

Gagel, G. and Ross, H. (2023). *Howard Ross discusses unconscious bias and inclusion* [Podcast]. 30 June. Available at: https://open.spotify.com/episode/626OFIWKg1RBJr8CWFC6iq [Accessed: 2 May 2024].

Gallagher, C. M. (2022, April). *The construction industry: Characteristics of the employed, 2003–20*. Available at: https://www.bls.gov/spotlight/2022/the-construction-industry-labor-force-2003-to-2020/home.htm [Accessed: 2 May 2024].

Greenwald, A. G. and Banaji, M. R. (1995). "Implicit social cognition: Attitudes, self-esteem, and stereotypes," *Psychological Review*, 102(1), 4–27. https://doi.org/10.1037/0033-295x.102.1.4

'Harassment.' (2024). *Merriam-Webster Online Thesaurus*. Available at: https://www.merriam-webster.com/thesaurus/harassment [Accessed: 20 May 2024].

Hinz, L. (2023). Interview by Gretchen Gagel [Zoom], 6 December.

Houssais, O. (2023). Interview by Gretchen Gagel [Zoom], 16 November.

Inside Construction. (n.d.). *Dr Gretchen Gagel*. Available at: https://www.insideconstruction.com.au/construction-experts/dr-gretchen-gagel/ [Accessed: 20 May 2024].

Jackson, B. (2023). Interview by Gretchen Gagel [Zoom], 11 December.

Kegan, R., and Lahey, L. L. (2009). *Immunity to change: How to overcome it and unlock the potential in yourself and your organization.* Boston, MA: Harvard Business Review Press.

Neuscheler, K. (2023). Interview by Gretchen Gagel [Zoom], 13 December.

Nussmeier, B. (2023). Interview by Gretchen Gagel [Zoom], 10 November.

Petrillo, K. S. (2023). Interview by Gretchen Gagel [Zoom], 20 November.

Reilly, T. (2023). Interview by Gretchen Gagel [Zoom], 25 October.

Welp, M. (2016). *Four days to change: 12 radical habits to overcome bias and thrive in a diverse world.* Zurich: EqualVoice.

Wonder. (2022, June 1). *How much information does the human brain learn every day?* Available at: https://medium.com/@askwonder/how-much-information-does-the-human-brain-learn-every-day-92deaad459a6 [Accessed: 20 May 2024].

Part II

"Me" – Grounded Self-Leadership – Introduction

I believe leadership begins with a focus on ourselves for two reasons: we are the person we have the most control over, and leadership is grounded in our authentic understanding of (and belief in) ourselves. I am a fan of authentic leadership, which is described by some as a distinct leadership style and by others as a characteristic of leadership, because at its core are many characteristics of leadership I believe to be important (Avolio *et al.*, 2004).

> *"Although efforts to define this construct have only recently begun, currently, authentic leaders are defined as those who are deeply aware of how they think and behave and are perceived by others as being aware of their own and others' values/moral perspectives, knowledge, and strengths; aware of the context in which they operate; and who are confident, hopeful, optimistic, resilient, and of high moral character"* (Avolio, Luthans, and Walumbwa, 2004; Avolio *et al.*, 2004).

The person you have the most influence over is you. You understand your purpose and values. You control how you show up as a leader. You control how you react to other people and life's challenges. Before we can influence others, we need to understand what it looks like to show up as our authentic selves. Spending time with yourself can be challenging. We have moments where we think "Wow, I'm awesome." We also have moments of fear, self-doubt, or even frustration and anger with ourselves. I believe we each need to accept and love ourselves as who we are while we work to improve and treat ourselves with kindness for being human and making mistakes.

> *"The most important and the most challenging person you will ever have to lead is yourself"* (Stoner, 2018).

From my PhD studies, research, and life experiences, I have created what I call the **"Grounded Self-Leadership Model"** because I believe self-leadership is "grounded" in your personal purpose and values (see Figure II.1). Grounded Self-Leadership begins with an aspiration to be a great leader, to continue building your leadership skills throughout your life. Grounded Self-Leadership means leading with courage, confidence, and the humility of knowing you will not always get it right. Grounded Self-Leadership means both effectively using your brain to think critically and make decisions, as well as resting your brain to fuel mental wellbeing and creativity. It is also about understanding your personal leadership brand and adeptly telling your leadership story.

As you read through Part II of this book, I encourage you to think about your aspirations as a leader, your strengths in each of these areas, and how you might select one or two areas to focus on enhancing. There is a great leader in all of you!

Figure II.1. Grounded Self-Leadership Framework.

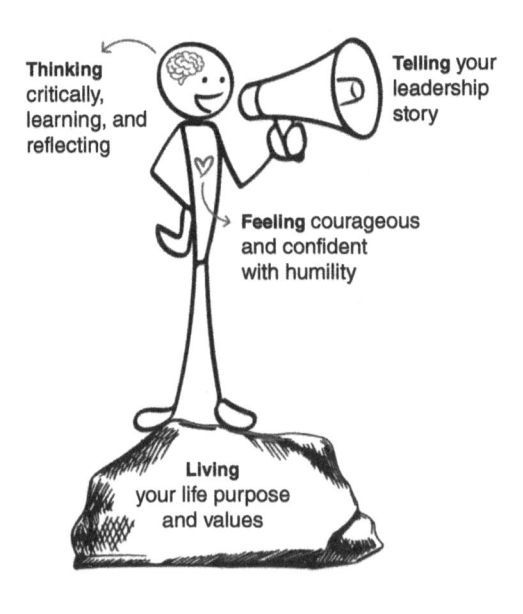

Thinking critically, learning, and reflecting

Telling your leadership story

Feeling courageous and confident with humility

Living your life purpose and values

References

Avolio, B. J., Luthans, F., and Walumbwa, F. O. (2004). *Authentic leadership: Theory-building for veritable sustained performance*. Lincoln, NE: Gallup Leadership Institute, University of Nebraska-Lincoln.

Avolio, B. J., *et al.* (2004). 'Unlocking the mask: A look at the process by which authentic leaders impact follower attitudes and behavior,' *The Leadership Quarterly*, 15, 801–823. Available at: https://doi.org/10.1016/j.leaqua.2004.09.003. [Accessed: 22 June 2024]

Stoner, D. S. (2018). 'The hardest person you will ever lead is yourself,' *Living Compass,* 22 June Available at: https://www.livingcompass.org/weekly-column/the-hardest-person-you-will-ever-lead-is-yourself [Accessed: 22 June 2024].

2

Defining Leadership and Your Leadership Aspirations

Introduction

Take a moment to reflect upon someone you admire, someone who you regularly go to for advice or assistance, someone you trust. Do you consider this person to be a great leader? Why? Leadership is not about a title; it is about character, behavior, skills, and values. My PhD studies provided me with an incredible opportunity to delve into 50 years of leadership theory. This, combined with my experience as the leader of multiple organizations and decades of coaching and developing leaders at every level, has given me a broad and deep perspective on important elements of great leadership. Great leaders come in all shapes and sizes, and different people lead in different ways. However, I do believe there are core characteristics of great leaders. I believe that great leaders are both born and created, that we are born with certain elements embedded in our DNA, and that we can continue honing our skills and behaviors as leaders throughout our lives. I also believe integrity and ethics are essential, as I have found that people do not follow people they do not trust.

> Friend and author Amy Jen Su shares this: "*Leader A, Gretchen, is like that part of all of ourselves where we wake up in the morning and somehow, no matter what challenge is thrown our way, we are swimming with the current. We're not taking things personally; we're just finding our own ease and effectiveness. Leader B is the contrast of that, where you wake up in the morning and somehow you know you're on the wrong side of the bed. Everything feels like an uphill battle. We're taking things a little more personally, maybe have less perspective, and it really elicits a more reactive side of ourselves. I do think it's a north star and aspiration. The goal isn't to be superhuman. The construct isn't meant to say, you must be perfect. It's*

> *an aspiration. It's the essence we know exists within ourselves. And really leader B is much more around. What are those conditions and factors where we lose access to the best of ourselves? The question I will often ask leaders is, what is your center of gravity, Leader A or Leader B? Do you have awareness of when you're in one mode versus the other? And if you do, do you have enough self-compassion to gently bring yourself back from B back to A."*
>
> (Amy Jen Su, speaking to Gagel, 2021 Greatness Podcast)

Thinking on leadership has evolved over time and will continue to evolve as we become more sophisticated in understanding our brains and human behavior. The command-and-control leadership style that worked well decades ago is not as effective in most contexts today. How you apply the elements of great leadership to fit your values—your personality—is critical. Great leaders also understand that the goal is not perfection; the goal is to lead, to learn, and to lead again, as discussed by Julia De Rosi, Deputy Assistant Commissioner of Project Delivery for the General Services Administration (GSA): *"Be the women you want to see; be the leader you wish you had. Have empathy and accountability, a great way to lead yourself and others. We are too hard on ourselves, our own worst critic"* (De Rosi, 2023).

> As Maja Rosenquist, Senior Vice President at Mortenson Construction, shares: *"My leadership style tends to be more masculine largely because that's what it took 20 years ago. The industry is changing, and women leaders need to embrace their authenticity, that not everyone needs to talk the loudest, be 'large and in charge.' It's okay to be collaborative, to 'listen to learn' versus 'listening to talk.' It's important that you figure out what your unique strengths are, what does executive presence look like that is authentic for you. Every woman leader needs to answer that question in a way that is authentic for them, that they are being true to their values."*
>
> (Maja Rosenquist, speaking to Gagel, 2021 Greatness Podcast)

The construction industry is full of great leaders solving the most complex challenges of society while humbly building the assets that support the existence of humankind. Again, leadership is not about a title; leadership involves influencing people—people looking to you for guidance and support. I know many people with important titles who are not leading, and many people with no official title or direct reports who are great leaders.

As you read this chapter, I encourage you to reflect upon where you are on your own personal leadership journey and "try on" different ideas to see if they are a fit for you. I believe that growing and developing as a leader is a lifetime journey—one of fun and exploration, trial and error, and good days and not so good days. To lead is to have courage, and I see a great leader in all of you!

Defining Leadership

When I ask people to consider the great leaders they have observed in their lives (like a coworker, public figure, or family member) and list the qualities that make those people great leaders, their thoughts often mirror the elements of great leadership documented in leadership research and theory. We intuitively understand many of the characteristics of great leadership, such as the skills and behaviors that define a person as a great leader, such as empathy, integrity, emotional intelligence, and the ability to motivate others. While many definitions of leadership exist, James McGregor Burns's seminal book *Leadership* contains many great thoughts (Burns, 1978).

> *"Leadership over human beings is exercised when persons with certain motives and purposes mobilize, in competition or conflict with others, institutional, political, psychological, and other resources so as to arouse, engage, and satisfy the motives of followers...in order to realize goals mutually held by both leaders and followers."*
>
> (Burns, 1978, p. 18)

I consider this quote to have many relevant points:

- **Leadership is about achieving a purpose.** We, as individuals, teams, business units, departments, and organizations, have SOMETHING we are trying to accomplish—"...persons with certain motives and purposes..."—such as gaining market share, executing safe construction projects, or satisfying customers (Burns, 1978, p. 18). We are all pointed toward an ever-shifting finish line of goals we are working to accomplish that enable the achievement of our collective purpose.
- **Leadership is about mobilizing resources.** To achieve these purposes, we mobilize different types of resources, "...institutional, political, psychological, and other resources..." (Burns, 1978, p. 18). We need these resources, deployed in the most effective way, to achieve our purpose.
- **Leadership is about motivation.** Great leaders realize that people do things because THEY are motivated to do it, not because we as leaders are motivated for them to do it. As leaders, we "...arouse, engage, and satisfy the motives of followers..." and must realize that different people are motivated by different things. That is one aspect that makes leadership challenging... and fun! (Burns, 1978, p. 18).

- **Leadership is about alignment on goals.** I have often heard leaders say, "My people aren't motivated to perform." One of the first questions I will ask them is, "Have they bought into the goals you've set for them?" Great leaders understand that effort is put forth "…in order to realize goals mutually held by both leaders and followers" (Burns, 1978, p. 18). It is the alignment on the finish line that is important.

Another favorite definition of leadership is that of my friend and mentor Edgar Schein and his writing partner and son Peter Schein, as described in *Humble Leadership* (Schein and Schein, 2023). *"Our definitions are important, and we all use the word leadership, but we don't really define what exactly does that mean and how does it differ from managing or other concepts of being the head of something. Our definition, which we're quite serious about, is that leadership is when someone implements something that is new and better. It must be new and better. A manager with a group can find a group member saying, 'You know, boss, we could do this in a different way,' The boss would say, 'That's a great idea. We're going to adopt that.' Both the boss and that member who suggested it have displayed leadership and the leadership is measured by the fact that the group is now going to be more effective in what it's doing."*

(Edgar Schein and Peter Schein, speaking to Gagel, 2019 Greatness Podcast)

I could share hundreds of additional definitions of leadership, and there are many overlapping elements in each, as well as unique thinking. I encourage you to create your own definition of leadership, as this is an important step in answering the question, "What does great leadership mean to me?"

Activity:

- Take a moment to think about additional definitions of leadership you have come across.
- How would you define leadership? What do you believe are some of the skills and behaviors of great leaders?

Leadership Styles

I often hear people describe leaders by their leadership style, such as a command-and-control style that loves to direct others or a laissez-faire leadership style that provides a great deal of autonomy to others. We use these mental models of leadership to help us quickly describe certain types of leaders and understand the elements of leadership. People's style of leadership can be specific to a context and can morph over time. One of my more challenging client engagements involved helping a seasoned CEO modify his style from the command-and-control style that had been necessary when he joined the organization to avert crisis to a more collaborative leadership style that was now necessary as the organization began empowering senior leadership. Adapting your leadership style to the context can be challenging, as this leader learned.

While many leadership style frameworks exist, these are three of my favorites:

Servant Leadership – First suggested by Greenleaf in a 1970 essay, "servant leadership" refers to a leader who believes they serve the needs of those they lead (Robert K. Greenleaf Center for Servant Leadership, n.d.). Servant leaders take time to understand the needs of the organization and those they lead to ensure that their efforts meet these needs. When I was hired as the President of an organization, I sat down with my team and asked questions such as, "If this organization is to be successful, what is the highest and best use of my time?" and, "What do you all need from me as a leader to be successful?" This demonstrated servant leadership.

Values-Based Leadership – James O'Toole writes of values-based leadership in his book *Leading Change: The Argument for Values-Based Leadership* (O'Toole, 1996). This leadership style is rooted in moral values and places a strong emphasis on respect. I believe great leadership is rooted in understanding and authentically conveying the values you stand for as a leader. We may place a different level of emphasis on certain values (such as collaboration or innovation), but those we lead understand that our leadership is grounded in ethics and integrity.

Transformational Leadership – This style, initially conceived by Burns and then expanded upon by Bernard M. Bass and Ronald E. Riggio in *Transformational Leadership,* focuses on leaders' ability to "transform" those they lead through four important factors (Burns, 1978; Bass and Riggio, 2005):

- Idealized influence – Influencing others as a leader because of the respect you command through your behaviors.
- Individualized consideration – Treating each person as a unique individual with specific needs and acting as a mentor and coach.
- Inspirational motivation – Challenging others to achieve high expectations, and motivating and inspiring those you influence to act.
- Intellectual stimulation – Encouraging others to be innovative and challenge their assumptions.

These are just three of the multitude of leadership frameworks that exist in the world. I think of these leadership frameworks as sets of clothing to try on as you create the unique leadership style that works best for you—a recipe where you add elements from each to make your perfect leadership dish. I encourage you to think about which elements of these leadership styles are most suited to you and worth cultivating.

Activity:

- Take a moment to consider these different leadership styles.
- Which of these seem to be natural styles for you?
- How would you describe your leadership style using elements of each of these?

Leadership Versus Management

For most of us, when we graduate from high school or college, our first position is that of a subject matter expert. We are doing the work we were trained to do as a carpenter, electrician, engineer, project manager, or accountant. At some point in our career, we may step into a management role and be taught management skills such as how to budget, allocate resources, and plan maintenance shutdowns. Whether your primary focus is on being a subject matter expert or a manager, you play a leadership role as well—either formally or informally—because leading is about influencing. Each of us can influence our teammates, our organization, and the culture of our industry, and I believe it is important to understand the difference between management and leadership as a starting point.

Dr. Brendan Nelson, regarded Australian political leader, Senior Vice President of The Boeing Company, and President of Boeing Global, shares this: *"What differentiates leadership from management in a civilian context is vision. You have to have a vision for your organization which inspires people to rise above themselves in their own self-interest, to be prepared to make sacrifices for the broader interests of the organization. Good leaders are people that are clear-minded and ethical as a matter of course, decisive, they're motivated by a sense of duty, a sense of the greater good for the organization and for others. You should always take your work seriously but never yourself. Good leaders in my experience are people that are witty, they're humorous, they work hard."*

(Brendan Nelson, speaking to Gagel, 2019 Greatness Podcast)

For decades, I have used the simple acronym "POC" (Planning, Organizing, and Controlling) to remember the fundamentals of management, and "SAM" (Set Direction, Align, and Motivate) to remember the fundamentals of leadership—both stemming from the work of John Kotter (2012) . Much of what I was taught early in my career at Ralston Purina was about the management skills that are critical to the success of an organization, as these create order, consistency, and results:

> - **P – Planning and Budgeting** – What is our plan, and how are we going to deploy resources to achieve that plan?
> - **O – Organizing and Staffing** – How will we best organize to accomplish the work? Who will fill what role, and how will we staff the team to leverage people's skills?
> - **C – Controlling and Solving** – What controls and systems will be put in place to ensure efficiency and accuracy? How will we go about solving complex problems?

What was eye-opening for me as I progressed in my career was the SAM of leadership—Setting Direction, Aligning People, and Motivating and Inspiring. Although I did not understand the difference between management and leadership at the time, I was learning leadership intuitively by observing other leaders and emulating what seemed to work. Many of Kotter's thoughts on leadership align with Burns' definition (Burns, 1978; Kotter, 2012).

> - **S – Set Direction** – As Burns states, what is our purpose? Where are we going? What are we trying to accomplish, and how do we measure success? People want to be on a winning team, and they want to understand what "winning" means.
> - **A – Align** – Burns' definition includes "...to realize goals mutually held...," an alignment on what we are hoping to achieve (Burns, 1978). This is a critical aspect of leadership that I believe often does not receive the attention it deserves. We assume the team is on board and aligned with what we are doing and how we are doing it.
> - **M – Motivate** – Again, back to Burns, "...satisfy the motives of followers..." (Burns, 1978). People are motivated by different things—both intrinsic rewards like job satisfaction and extrinsic rewards like bonuses. Often, when I ask leaders about the unique motivators of each member of their team, these leaders lack clarity on the topic.

These leadership skills and behaviors create movement and change. I believe that being able to differentiate between management and leadership is critical because many people THINK they are leading when they are in fact managing. Both are important, and being able to differentiate between management and leadership allows one to focus on the skills required for each to ensure that there is adequate focus on both management and leadership.

> As team expert Michelle Terkelsen shares: *"We call it a diary audit where we get people to audit their diary over a course of a week or two weeks and bucket where they are spending their time. They rate the importance of meetings for example. They start to think, wow, I really need to change things up and put more time into this bucket called leadership."*
>
> (Jan Terkelsen and Michelle Terkelsen, speaking to Gagel, 2020 Greatness Podcast)

I frequently ask leaders to draw a circle on a piece of paper, and then write down at the top of the circle how many hours they typically work in a week. Next, I ask that leader to divide the circle into three slices: one for the time they typically spend as a subject matter expert (doer), one for the time they typically spend as a manager, and one for the time they typically spend as a leader. I point out that every week is different and that this is not a statistical exercise, just an approximation! I also emphasize that there is no "correct" answer. Every person's role is different and requires a different mix of time as subject matter expert (doer), manager, and leader (Figure 2.1).

Figure 2.1 Example: Current Allocation of Time.

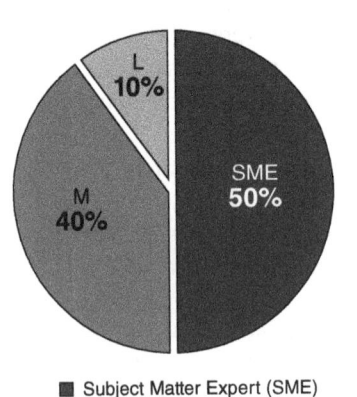

■ Subject Matter Expert (SME)
■ Manager (M)
■ Leader (L)

I then ask each leader to draw a second circle with the same number of hours at the top. Then, I ask them to draw a slice for how much time they SHOULD be spending as a subject matter expert, how much time they SHOULD be spending as a manager, and how much time they SHOULD be spending as a leader to ideally execute their role within the organization (Figure 2.2).

Figure 2.2 Example: Ideal Allocation of Time.

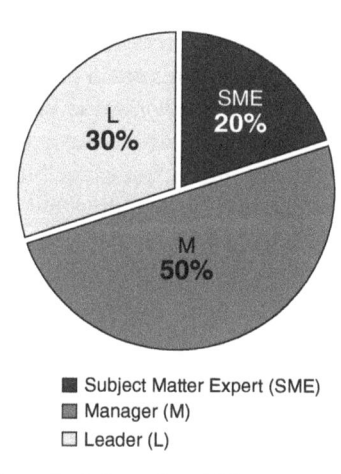

■ Subject Matter Expert (SME)
▦ Manager (M)
☐ Leader (L)

When I ask a group of leaders how many of them feel they should be spending more time as a subject matter expert, one or two hands go up. When I ask the group how many of them feel they should be spending more time on management activities, one or two more hands go up. When I ask the group how many of them feel they should be spending more time leading, nearly all hands go up, and these results are consistent across many groups. I often ask these groups, "What are the benefits of spending time on leadership?" The response is immediate and affirming: that there is a significant benefit to spending time on leadership because people are more motivated, aligned in what activities are a priority, and more engaged in the work of the team.

If many of us feel that we should be spending more time on leadership, what are the barriers to doing so? When discussing this with leaders, I hear things such as:

- "I am too buried in the tasks of subject matter expert or manager to take time to lead."
- "My organization does not value leadership, and they value getting things done."
- "I don't really know what I'm supposed to be doing to lead."
- "Leadership feels too 'squishy,' like I'm not really doing anything."
- "I'm too busy to delegate."

When we start out as subject matter experts, we are probably doing something we were given specific training to do—including learning how to do it correctly. When we complete a task, we typically know if it was done correctly or incorrectly, and we feel pride in doing a job well done. For example, if you create an engineering or architectural drawing, it is reviewed, corrected, and finished, and you can check that box. If you are a carpenter, you know if the framing has been done correctly.

As we progress to management duties, we are again primarily doing tasks we were trained to do. We might be a project manager on a construction project, assigning resources to tasks to ensure we stay on schedule and making sure that safety procedures are in place and adhered to. We might be an electrical foreman overseeing the work of several electricians, ensuring we are scheduling the work and that everyone has the materials they need to do the work.

Leadership is "squishier." It's hard to check a box that a specific task has been completed. Leadership is not about a title or a reporting structure; leadership is about people looking up to you and your ability to influence people and implement change. Leadership is critical. But who trains us in the skills and behaviors necessary for great leadership? Who teaches us how to listen? How to have empathy? How to motivate different people who are motivated by different things? Although we are investing more in leadership development as an industry, I would argue that it's still not enough. Even when we train people on what to do to lead, the pressures of completing tasks as a subject matter expert or manager often take up most of our time. **Leadership takes time.** Leadership requires thoughtful reflection, and we often don't receive immediate feedback on whether it's "right" or not like we would as a subject matter expert or manager.

> *"When we're subject matters experts or technical experts or practitioners we've been rewarded for a long time for being the one driving the ball down the field. And now suddenly as a manager, leader, you're being asked to also play coach. Redefining that ratio of player to coach is really critical."*
>
> (Amy Jen Su, speaking to Gagel, 2021 Greatness Podcast)

Creating the space to reflect is one of the most important things a leader can do, and having the discipline to create boundaries around that time is essential. As CEO/President of a management consulting firm, I was flying about 120 flights a year, and I almost never connected to Wi-Fi. Why? Because this was prime leadership time. How could I be leading while separated from my team on a plane and not connected to Wi-Fi, you ask? Because I was thinking. I was thinking about how we were doing on our strategic plan. I was thinking about the people on the team and how they were feeling about things. I was thinking about what the next important conversation I had to have with people during their quarterly meetings was. Time to reflect is critical as a leader.

> Frequently, the only way to free up time for leadership is to delegate subject matter expert and manager tasks, and this can be challenging for many of us. Do you sometimes feel that it's just faster to do it yourself? Are you nervous that if you delegate a task, it will not be completed to your standards? These

> are common statements I hear. Your ability to effectively delegate is a critical component of your ability to lead, and if this is a challenge for you, I recommend you spend additional time exploring resources to improve your ability to delegate.

Taking the time to think about management versus leadership, understanding the difference between the two and how your time should be allocated, form a critical component of self-leadership. Leadership will not happen if you do not devote time to it, and the benefits to you and your team are tremendous.

Activity:

- Draw a circle and divide it into three segments representing the percentage of time you currently spend as a subject matter expert (SME), versus manager, versus leader.
- Draw a second circle and divide it into three segments representing the percentage of time you SHOULD spend as a subject matter expert (SME), versus manager, versus leader.
- Reflect upon the differences.
- What one thing could you do to shift more time to an area that requires more time?

Your Leadership Aspiration

I was conducting a performance review of a senior leader in the construction industry, when I had a bit of a "light bulb" moment. After sharing specific feedback (both positive and areas for improvement), I looked at this leader and said something to the effect of, "You are at a pivotal point in your career. You are a 'good' leader. I believe you know what it takes to become a 'great' leader. The question is whether you will invest energy into developing the skills that will take you to this next level of leadership."

This is an important question for you: ***Do you aspire to be a great leader?***

I do NOT aspire to be a great golfer. I do not keep a handicap, and I spend just as much time taking pictures of birds, animals, and trees while I'm golfing (and avoiding poisonous snakes while in Australia!) as I do hitting golf shots. If I aspired to be a great golfer, I would invest in much-needed lessons and spend time practicing at the driving range. I don't do these things because being an "okay" golfer who has fun and enjoys nature is my end goal. I do not aspire to be a great golfer.

I ask you to deeply contemplate this question: **Do I aspire to be a great leader?** It's okay if the answer is "no," if you want to be a great subject matter expert or manager, or if you want to be an "okay" leader. You will still benefit from this book. If you DO aspire to be a "great" leader, you will continue to invest in yourself as a leader, learn, build new skills, and measure your progress; not necessarily progress measured by promotions, but progress in your ability to influence people and command respect and admiration.

Aspiring to be a "great" leader will cause you to:

- Deeply reflect on your strengths and opportunities.
- Invest in your personal development and possess a growth mindset.
- Create feedback loops on your performance as a leader.

Another important question for you: ***Why do I aspire to be a great leader?*** Is it for the accolades? The titles? The accomplishments? To serve a greater purpose? To serve your team and/or organization? I ask this because we are going to spend a great deal of time reflecting upon how you can build others' trust in you, as well as how you can trust others. People do not follow leaders they do not trust, and people do not follow leaders who are in it solely for their own glory. That does not mean we cannot be excited about that new promotion or the positive feedback we receive. One of my languages of love is words of appreciation—I love to be appreciated! (Chapman and White, 2019). It means that glory is not the primary reason we lead. We lead because we care about people, we care about our teams, we care about our organizations, and we care about our industries.

Being clear on your aspirations as a leader is important because it defines the energy you are willing to commit to continuing to develop as a leader. Just like golf, leadership takes practice, training, coaching, and learning. Being an effective leader is gratifying—especially as you develop new skills and reap the rewards of stronger relationships and influence.

Activity:

- Take a few minutes to reflect upon your aspirations as a leader. Where are you today compared to where you'd like to be? Do you aspire to be a great leader?
- Think about yourself 10 years from now. What are people saying about you as a leader? What skills might you need to build to cause them to say this about you?

Conclusion

Skills as a subject matter expert, as a manager, and as a leader—all three are important. Dedicating the appropriate time to each, given your role in the organization, is critical. Ensuring that you are spending enough time on leadership (and we'll learn more about what that looks like throughout this book) is vital if you are going to achieve your leadership aspirations.

I encourage you to continue honing your own personal definition of what it means to be a great leader, think about the elements of great leadership as an inventory of skills, and take the time to reflect on your strengths and what skills might need further development. You will never be a "10" on everything. Leverage your strengths and pick one or two skills from the topics discussed in this book to work on—skills you feel will have the greatest impact on your leadership capabilities. For me, that has been to be more present, be a better listener, and have great emotional intelligence and empathy. You will have your own list, and I'm confident that this book will help you understand where to focus your efforts.

> *"Embrace your vulnerable edge. It took me a long time to understand the importance of vulnerability. As my career progressed, I found myself losing that vulnerability and trying to mirror my male counterparts. Luckily, I realized that vulnerability is not a weakness, if anything, it is a strength and a differentiator. It's important to be genuine, be comfortable with who you are, and never feel nervous or ashamed."*
>
> (Petrillo, 2023)

Different people will lead in different ways, and that is okay. Great leadership is a gift you give to those around you and to yourself. Given the complexity and volatility of today's world and the construction industry, people seek out leaders who will help assure them of the path forward and how to achieve success. Whatever your aspirations are, however far you would like to develop your leadership skills, I see a great leader in all of you!

References

Bass, B. M. and Riggio, R. E. (2005). *Transformational leadership*. New York: Psychology Press.

Burns, J. M. (1978). *Leadership*. New York: Harper & Row Publishers.

Chapman, G. and White, P. (2019). *The 5 languages of appreciation in the workplace: Empowering organizations by encouraging people*. Woodmere, NY: Northfield Publishing.

De Rosi, J. (2023). Interview by Gretchen Gagel [Zoom], 20 November.

Gagel, G. and Nelson, B. (2019). *Dr. Brendan Nelson discusses leadership and personal values* [Podcast]. 10 May. Available at: https://open.spotify.com/episode/ 7cadZRiknfvUF63YeWcAlj [Accessed: 28 April 2024].

Gagel, G. and Rosenquist, M. (2021). *Maja Rosenquist discusses diversity and leadership in the construction industry* [Podcast]. 3 May. Available at: https://open .spotify.com/episode/7N1ZEIpLVNlt9kQnzCIDe0 [Accessed: 28 April 2024].

Gagel, G., Schein, E., and Schein, P. (2019). *Ed & Peter Schein discuss their book Humble Leadership* [Podcast]. 24 June. Available at: https://open.spotify.com/ episode/5JueFC2LHdo74Ncz4HqVs5 [Accessed: 28 April 2024].

Gagel, G. and Su, A. J. (2021). *Amy Jen Su discusses her book The Leader You Want to Be* [Podcast]. 12 November. Available at: https://open.spotify.com/episode/ 4ePtBXmgN40IG7S7pKiFu6 [Accessed: 1 May 2024].

Gagel, G., Terkelsen, J., and Terkelsen, M. (2020). *Jan and Michelle Terkelsen discuss creating high performance teams* [Podcast]. 3 February. Available at: https://open .spotify.com/episode/4U22Jcl7lYP2Ana37I4zrI [Accessed: 28 April 2024].

Kotter, J. P. (2012). *Leading change*. Boston: Harvard Business School Press.

O'Toole, J. (1996). *Leading change: The argument for values-based leadership.* New York: Ballantine Books.

Petrillo, K. S. (2023). Interview by Gretchen Gagel [Zoom], 20 November.

Robert, K. Greenleaf Center for Servant Leadership (n.d.). *What is servant leadership?* South Orange, NJ: Greenleaf Center for Servant Leadership. Available at: https:// www.greenleaf.org/what-is-servant-leadership/ [Accessed: 26 May 2024]

Schein, E. H. and Schein, P. A. (2023). *Humble leadership: The power of relationships, openness, and trust.* 2nd edn. Oakland: Berrett-Koehler Publishers.

3

Living Your Life Purpose and Value

Introduction

Three months before my planned move to Australia, I attended the International Women's Forum (IWF) annual conference in Melbourne, Australia. I joke now that the Australians were probably wondering, "Who is this weird woman from Denver walking around saying, 'I'm moving to Australia, will you be my friend?!'" Out of the many amazing women I met, Dr. Sue English stood out to me. When I moved to Australia three months later, I lived in Canberra, Sue lived in Melbourne, and we occasionally saw each other at IWF events. Then, on April 6, 2020, I moved by myself to Melbourne for a new position, and right into the COVID lockdowns. I was living alone in a high-rise apartment with few friends, and it was probably one of the loneliest times of my life. Sue sensed this and convinced a friend who had organized a weekly virtual gathering for a dozen of her friends to include me. These gatherings became my lifeline through six lockdowns, totaling 262 days.

Sue exemplifies leadership not because she has a big, fancy title or leads hundreds of people, but because I know exactly what Sue's purpose is—to be a wonderful friend, mother, and medical doctor; to help those in need, to sense when someone is struggling, and to be there for them. I know what Sue values: family, deep relationships, integrity, kindness, and her own health. I know Sue's personal brand: impeccable host, reliable friend, and pillar of the community. Yes, Sue sits on boards, was State Committee Chair Victoria for IWF Australia, and has been awarded an Order of Australia (OAM). But it is Sue's informal leadership within her family, friends, and the medical community that causes me to admire her as a great leader.

> As friend, author, and speaker Janine Garner shares: *"The first law is around owning your spotlight. It's about really understanding who it is that you are, who it is that you are being, and who it is that you want to become; because if you can't own that stuff, if you can't own your own story, your own narrative, your own values, your own dreams, the things that you're good at, the things that you're rubbish at, then you're just faking it till you make it and I totally disagree with that line of thinking. Simon Sinek talks about identifying your 'why' personally and I think until you can do that work of really understanding who you are, owning that narrative, owning that backstory, owning the good stuff, owning the warts, owning your dreams, you're limited as a leader. The first law is about owning your spotlights and really stepping into that place of courage, of confidence, of conviction, of being able to share with the world who you are."*
>
> (Janine Garner, speaking to Gagel, 2022 Greatness Podcast)

When I think of the many great leaders I have advised, their purposes and values are clear to me. This is the foundation of Grounded Self-Leadership: to deeply understand yourself as a unique and authentic person, and to bring that to your leadership. Great leadership starts with what you stand for as a person, a leader, a follower, and a teammate.

Throughout this chapter, I will ask you to deeply reflect upon yourself. This could take days, weeks, or months, and it's not a "one and done" process. We are a work in progress throughout our lives, and you will probably revisit these exercises periodically. As you read through this chapter, I encourage you to invest deeply in this reflection—you are worth it!

Personal Purpose

I first read the book *The 7 Habits of Highly Effective People* by Stephen R. Covey (2020) during my first career as a manufacturing operations manager. In the book, Covey asks you to envision yourself at your funeral. What are people saying about you? What stories are they telling? Then, he asks you to think about what you WANT them to say about you at your funeral. Is it the same? Is it different? This was a deeply meaningful exercise for me because it was the first time I purposefully reflected upon what I cared about most and why I existed.

It is easy in life to go down paths because we are doing what is expected of us based on societal norms, family expectations, or our own voice telling us what we should or should not do. I think of life as a series of forks in the road that are the many choices we make (none of them right or wrong) which impact our lives—choices like where we go to school, what career we choose, or who we marry. I believe it is helpful to stop periodically and reflect upon the intentionality of living the life we choose.

> As Josephine Sukkar, Founder and Principal of Australia's Buildcorp and Independent Director on numerous boards, states: "*Follow your heart, do what you do. Only you know your personal circumstances.*"
>
> (Sukkar, 2023)

Purpose is defined as "the reason for which something is done or created or for which something exists" ('Purpose,' 2024). "Why do I exist? Why am I here? What am I supposed to be accomplishing on this planet?" These are deep questions that can overwhelm us and to which there is no right answer, but a very personal answer for each of us. After I went through the funeral exercise from Covey's book, I determined that my purpose in life was **"to leave the world a better place than I found it"**—a statement that guides my decisions in life. For example, when I was approached about writing this book, I thought to myself, "Will this leave the world a better place?" Yes, if I can help the construction industry be more inclusive, help women thrive in the industry, and help men understand how to better support them, then "yes." As friend, author, and life coach Christy Belz shares: "*Each moment is a choice point to choose again. My book's really about the tools and techniques supported by stories to help us get more present, get us more centered in our lives and get us more in a place of actually being in our lives instead of doing, doing, doing; being the human **beings** instead of the human **doers**"* (Belz, 2022; Christy Belz, speaking to Gagel, 2022 Greatness Podcast).

> Prior to moving to Australia, I decided to carry a couple of copies of my earlier book, *8 Steps to Being a Great Working Mom,* with me each week to give away to working moms (Gagel, 2015). One night, I was on a Southwest Airlines flight and one of the flight attendants had a large "Great Mom" button on her apron. I brought a copy of the book up to her in the galley and said, "I bet you are a great working mom," and handed her the book. She burst into tears, explaining to me that she worked as a flight attendant on the weekends to pay for her kids' sports, but that it meant she was not with them on the weekends, and she felt tremendous guilt about that. My telling her she was a "great mom" meant something to her. To me, that is leaving the world a better place than I found it, as it made her feel grateful for my acknowledgment of her support for her children and the sacrifices she made, and it made me feel good because I had acknowledged someone's worth with a random act of kindness.

Sometimes, we may not clearly understand our purpose. Coach, author, and speaker Christelle Pillot shares how she herself was suffering from depression in her job as an engineer and realized through deep reflection that this was not her purpose in life: *"The first thing is to know is who you are, your strengths and your skills. The second thing is to look at, what energizes you? What excites you? What engages you? Not only at work, look everywhere. What kind of TV do you like to watch and why do you like to watch that? I love energy management because energy management shows us a little bit of where we want to put our energy and where we get energy back. You look at your day or you look at your last year. When did I feel really alive?"* (Christelle Pillot, speaking to Gagel, 2023 Greatness Podcast). The term "flow" was popularized in part by the book *Flow: The Psychology of Optimal Experience* by Mihaly Csikszentmihalyi (2008) and describes a state where we feel "in the groove" (Csikszentmihalyi, 2008). Time passes without us noticing it. We look up thinking, "Wow, that hour went fast," or, "That day went fast." I believe that when we are living with purpose—when our actions come from a place of deep authenticity—this is the energy we feel.

> I respect the two-by-two quadrant that friend and author Amy Su shares in her book, *Be the Leader You Want to Be* (Su, 2019). *"The 2 by 2 Purpose Quadrant really holds two criteria, two elements. Our purpose is a function of both our contribution, so at any given point in your life, what is the highest and best impact contribution you hope to make for the world, for the organization, for your family; and at the same time our purpose is ever evolving and a function of our passion, what gives you energy today. If you found yourself with a free hour, what would you find yourself working on, what are you drawn to? As you begin to walk through your life looking for the intersection of those two things, then you end up in the upper right-hand quadrant where you have both high contribution and really high passion; and to me that's a real foundational point of living as a leader where, wow, how great is it that the difference we're trying to make or the contribution we're having in our organization happens to also be what we enjoy doing and are passionate about?"*
>
> (Amy Jen Su, speaking to Gagel, 2021 Greatness Podcast)

I encourage you to regularly invest the time you need to reflect about yourself, what you care about, and your purpose in life. This is one of the greatest gifts you can give yourself. Your purpose is the foundation of being an authentic leader and requires tremendous vulnerability, and it may change over time because what you care about is influenced by your life experiences. Taking time periodically to reflect upon your purpose in life, being confident in understanding the contribution you are meant to make to your team, your organization, your industry, your family, and living with passion for what you believe in is one important foundational element to being a great leader.

> **Activity:**
>
> - If people were to attend your funeral today, what would they say about you? What would you WANT them to say about you? Are these statements consistent? What differences exist, and why?
> - Reflect upon your purpose in life. Why do you exist? How do you make the world a better place? Is that done through your work, your children, or your community?
> - What do you want people to say you have accomplished in your life, and why?

Personal Values

It was about 9:00 a.m. on an ordinary morning in 2009. I was having a one-on-one meeting with Meg Ferland, then-President of the Colorado Children's Campaign, in her office in Denver. The irregular heartbeat I had been trying to ignore for weeks was becoming dramatically worse that morning. I tried to calm my heart, calm myself, and ignore it, but nothing was working. Finally, I said to Meg, "I'm really sorry, but I think I need to go see my doctor." After an EKG, a three-day Holter Monitor test, a stress test, and an ultrasound, it was determined that I was having premature ventricular contractions (PVC's) and premature atrial contractions (PAC's). Essentially, all four chambers of my heart were beating out of rhythm. What this felt like to me was "beat, beat, long pause... beat"—quite disconcerting! My doctor said it was as if the coating had come off one of the "wires" on my heart, causing it to send its signal at the wrong time. The most probable cause was stress and not taking care of myself.

At the time, I valued everything and everybody else above me—my kids, my partner, my work, my volunteer work. I was not exercising or doing ANYTHING to take care of myself. I am grateful that I suffered from a health care problem I could recover from versus just dropping dead of a heart attack. It was a huge wake-up call and caused me to start valuing myself. I began doing yoga and eventually progressed to the gym Orangetheory, where I have completed over 700 sessions (Orangetheory Fitness, n.d.)! I gave up caffeine and started sleeping seven-to-eight hours a night. After five years, I was able to stop taking my beta blocker, and now I only feel these PVC's and PAC's when I am overly tired or stressed—a great signal to slow down and take better care of myself. All of this caused me to value my health in a way I had never done before.

> *"I would say that 95% of the people that come into my world are exhausted. Leaders are exhausted, business owners are exhausted, parents are exhausted, friends are exhausted. We're exhausted mentally. We are suffering from so much overload that we can't see the woods for the trees. I ask, 'Do you have time right now where you've got nothing to do?' We need that time. Because if you're exhausted, if you are burnt out, if you can't think clearly, if you cannot get perspective, if you cannot be curious because you're rushing around trying to make decisions and trying to get projects done, you're not bringing the best of yourself to your work."*
>
> (Janine Garner, speaking to Gagel, 2022 Greatness Podcast)

A value is defined as "… something (such as a principle or quality) intrinsically valuable or desirable" ('Value,' 2024). Just as people have unique purposes, different people value different things in life. Some of us place more value on material things, and some on experiences. Some of us value trying new things, and some of us value stability and routine. Your values are core to who you are as a human being and as a leader, and (along with your purpose) are your personal compass. Values guide your decisions and help you navigate a complex world of choices, as well as help define your Grounded Self-Leadership. Taking time to write down your values is a great starting point. My good friend and emotional intelligence coach Brent Darnell led me through a process of brainstorming all my values, then cutting the list to 10 values, and then cutting it down further to five values. This is an interesting exercise because it forces you to focus on the values that are most important to you—the non-negotiable values that define you.

> *"I believe that you must be really clear about what your values are, and you have to be willing to stand for those values in the face of adversity and conflict. My values are love. My values are connection. My values are collaboration. I look for those places to make a difference. Knowing your values and espousing them and then being willing to stand for those values even in difficult times is my personal belief of what makes leaders terrific."*
>
> (Christy Belz, speaking to Gagel, 2022 Greatness Podcast)

Sometimes, values will clash. I value my family and I value my clients. I have time scheduled with my family, but then a client has a crisis—so what do I do? There is no "right" answer, and the important thing is that you make the decision that is right for you. Joseph L. Badaracco (2016) describes these as "defining moments" and utilizes various scenarios to illustrate the "right versus right" decisions we make based upon our values. For example, Badaracco (2016) writes of a young consultant who is chosen to join a final report-out to a client because he is Black, even though he has not been involved in the work to date. This consultant struggles because he values being a team player, but also does not value being placed in a tokenistic position. He reaches a decision to participate in the presentation only if he is brought in to complete the work with the team, thus satisfying both values. There is no right or wrong decision—only the decision that best aligns with our values.

Years ago, I started doing an exercise I call the "Pie of Life." I draw a circle in my journal and write "168" at the top—the number of hours we have in a week (and by the way, we only have about 4,000 weeks on average in our life; use them wisely! [Burkeman, 2021]). Then, I carve the circle into pie slices of time that represent how much time on average I want to spend on different categories in my life based on my values and priorities. What "slice" do you think I start with first? Sleep! If I don't sleep, my heart starts beating irregularly—it's that simple. The next slice is exercise. I am no good to anyone if I am not alive—sleep and exercise are critical. My next slice is family, then friends… then work, my volunteerism, my teaching, my writing, my fun time, and golf (Figures 3.1 to 3.3)!

We often feel that we can "blow up the balloon" of life and make this pie bigger, but we cannot. We all have the same number of hours in a week. When something comes up that causes one piece of the pie to become bigger, that time must come from somewhere else. That is the juggling act of life, and prioritizing what we value is important.

Figure 3.1 Pie of Life: Example, Early Career.

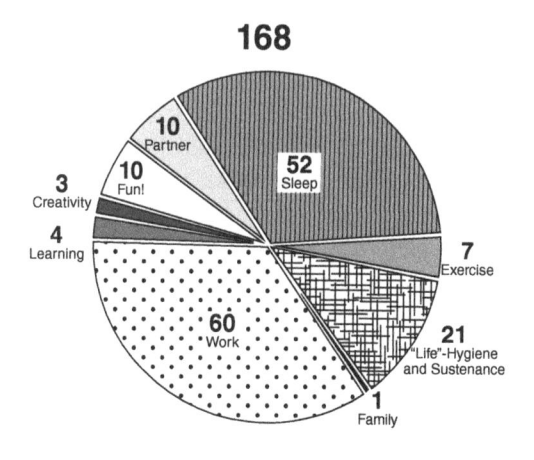

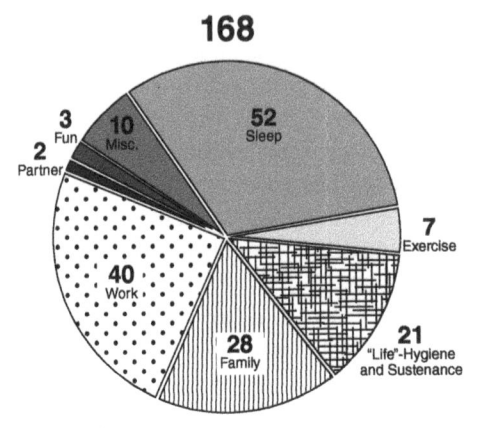

Figure 3.2 Pie of Life: Example, Mid-Career.

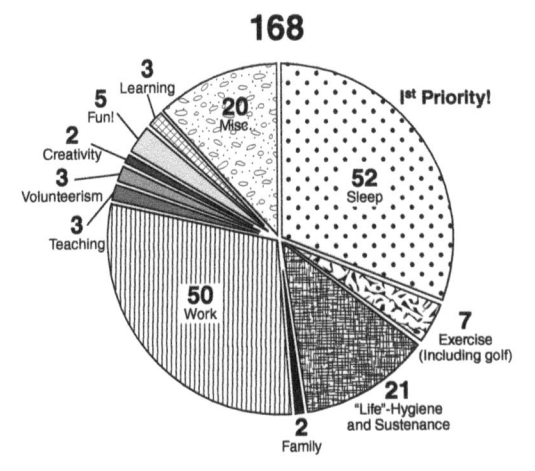

Figure 3.3 Pie of Life: Gretchen Example.

Our Pie of Life changes over time. In February 2013, my wonderful mother was in the last stages of Alzheimer's. The piece of the pie devoted to helping her became bigger and bigger with falls, emergency trips to the hospital, and time spent with her during her final days. My Pie of Life felt as though it would burst, given my responsibility as a mom to two teenagers and my work as assistant dean of the Daniels College of Business at the University of Denver. Thanks to the support of my friends and financial advisor, I made the decision to stop working for a few months, even though I was recently divorced and had two teenagers about to head off to college. I stepped down from my position in February 2013, my mom was in hospice by July, and she passed away in October. I am so grateful I had more time with her.

I am not advocating that everyone quit their job! But for me, this was the "right versus right" decision. If any piece of your pie is getting bigger—self-care, kids, aging parents, health challenges—it must come from somewhere. I am advocating that you take care of yourself and realize that just like everyone else, you have only 168 hours in a week. You cannot manufacture more. You must have the courage to make choices in how you spend your time that align with your values. You vote for your priorities (what you value) with your time.

Stephen R. Covey (2020) talks about our top priorities like the big rocks you should put in your jar first. Each of us has many small priorities that can take up big portions of time and fill up our jar first, making it hard to carve out time for our "big rocks"—our top priories. It is important for you as a leader to determine the big rocks for your personal life and your role as a leader and teammate—your most important priorities and highest contributions to the world. Next, you need to create boundaries. There is always going to be more to do on a project, one more team member to meet with, and one more conversation to have. You have the power to create boundaries, and it starts with a strong signal to those around you as to what you value as a leader. You have the power to put the big rocks in your jar first and ensure that you dedicate an appropriate amount of time to your top personal and work priorities (Figure 3.4).

Figure 3.4 Putting the Big Priorities in the Jar First.

Focusing on your values also involves "controlling the controllable," as Australian Short Track Cycling Olympic champion Anna Meares discussed when talking about her mindset after breaking her neck during a fall: "*It was pressured circumstances we found ourselves in with the shifting path that*

> *I now had to walk to make my goal of competing in the Beijing Olympic Games. I realized I needed to really be selective as to where I put my time, my energy, and even my money because I didn't have much of it and I didn't have much time either. I had to become very deliberate as to who I engaged with, why I engaged with them, and what the result of all those interactions were. I realised that there was a lot in life that I could not control, nor could I change the outcome of, so I started to focus a little bit backwards on what can I change the outcome of, and therefore what can I control in that scenario, and that's where I put my time my energy and even my money.*
>
> *"That's a big lesson for a 23-year-old which is why I think you can define my career to pre- and post-accident, because there were some significant learnings forced upon me at that time which I'm grateful for. I think it gave me a great change in perspective, a shift in my gratitude and my appreciation for life, what I was doing, who I was working with. It also allowed me to understand that I was far more capable than I realized prior to my accident. Many people don't fully comprehend their full capabilities until they've been put in a position where they've either hit rock bottom, or their back is up against a wall. This challenges the value of what it is they're chasing and why they're chasing it. For me to come back into a sport and a competitive environment where I had been so damaged because of a fall just reiterated to me the passion that I had for it. As a result, I won 9 of my eleven world titles after my accident. I had my most successful Commonwealth Games after my accident, and my biggest Olympic win after my accident as well. I think honestly none of that would have come without the lessons learnt in this adversity of my accident in 2008."*
>
> (Anna Meares, speaking to Gagel, 2020 Greatness Podcast)

Taking time to reflect on your unique values and how you communicate those values to others as a leader lets people know what you stand for. Clarity on your values not only makes you a great leader but also helps you find organizations and teams whose values align with yours. There will be times when it requires tremendous courage to stand up for your values. Your values, along with your purpose, are the foundation of who you are as a leader and a human being.

Activity:

- Take a moment to brainstorm your values. Now cut the list down to 10 items. Now cut the list to five.
- In two or three sentences, describe why each of these values is important to you and how you will be present to them in your life.

- What does your Pie of Life look like today? Does it reflect your values? What changes might you make to align your time with your values?
- Revisit these values periodically. Changes in your life (such as having kids, aging parents, or health challenges) continue to shape your values and the priority of each value.

Conclusion

Your values are an important component of the foundation you stand upon as a great leader. Your purpose and values are unique to you—they are the rod that goes right through your middle and allows you to stand tall and proud. You count. You make a difference as a leader. Your purpose and values are valid, and remaining true to them defines you as a great leader. People will appreciate understanding what you stand for as a leader and the courage it may require at times to be true to these values.

> *"Be true to yourself. Don't lose sight of who you are because when you start trying to please others, are trying to listen to criticisms and say, 'Oh, I need to react', you don't necessarily need to act, to react to anything. You can just be still. I always like to be still for a minute and then decide, okay, what part of that pertains to me? What part of that impacts me? And what part of that do I want to take on in the sense of a goal?"*
>
> (Denise Burgess, speaking to Gagel, 2023 Greatness Podcast)

References

Badaracco, J. L., Jr., (2016). *Defining moments: When managers must choose between right and right*. Boston: Harvard Business School Press.

Belz, C. (2022). *Oh god of second chances, here I am again*. New York: Guide Point North Publishing.

Burkeman, O. (2021). *Four thousand weeks: Time management for mortals*. New York: Farrar Strauss & Giroux.

Covey, S. R. (2020). *The 7 habits of highly effective people*. 30th anniversary edn. New York: Simon & Schuster.

Csikszentmihalyi, M. (2008). *Flow: The psychology of optimal experience*. New York: Harper.

Gagel, G. (2015). *8 steps to being a great working mom*. Atlanta: BDI Publishers.

Gagel, G. and Belz, C. (2022). *Christy Belz discusses her book Oh God of Second Chances* [Podcast]. 29 December. Available at: https://open.spotify.com/episode/2MGOy3gef2pHKAlx9LmGXq [Accessed: 1 May 2024].

Gagel, G. and Burgess, D. (2023). *Denise Burgess discusses her career as an African American woman leader in the construction industry* [Podcast]. 29 December. Available at: https://open.spotify.com/episode/77UCAi1FXUKeMOS6yfes8z [Accessed: 2 May 2024].

Gagel, G. and Garner, J. (2022). *Janine Garner discusses her book Be Brilliant* [Podcast]. 14 October. Available at: https://open.spotify.com/episode/2jTwgPFgU2jmhi9kG6WNTW [Accessed: 1 May 2024].

Gagel, G. and Meares, A. (2020). *Anna Meares discusses leadership and team lessons from being an Olympic athlete* [Podcast]. 17 February. Available at: https://open.spotify.com/episode/3k1RqB8K3aiLVG5lRgkRol [Accessed: 28 April 2024].

Gagel, G. and Pillot, C. (2023). *Christelle Pillot discusses finding a career you are passionate about* [Podcast]. 14 July. Available at: https://open.spotify.com/episode/5cBIkPTBrxLK1dPtNXNxhR [Accessed: 2 May 2024].

Gagel, G. and Su, A. J. (2021). *Amy Jen Su discusses her book The Leader You Want to Be* [Podcast]. 12 November. Available at: https://open.spotify.com/episode/4ePtBXmgN40IG7S7pKiFu6 [Accessed: 1 May 2024].

Orangetheory Fitness. (n.d.). *Experience the smarter 1-hour workout*. Available at: https://www.orangetheory.com/en-us [Accessed: 27 May 2024].

'Purpose.' (2024). *Merriam-Webster Online Dictionary*. Available at: https://www.merriam-webster.com/dictionary/purpose [Accessed: 27 May 2024].

Su, A. J. (2019). *The leader you want to be: Five essential principles for bringing out your best self—every day*. Boston: Harvard Business Press Review.

Sukkar, J. (2023). Interview by Gretchen Gagel [Zoom]. 21 December.

'Value.' (2024). *Merriam-Webster Online Dictionary*. Available at: https://www.merriam-webster.com/dictionary/value [Accessed: 27 May 2024].

4

Feeling Courageous and Confident with Humility

Introduction

The first time I walked into a manufacturing plant as an operations manager trainee, I was scared to death. I remember thinking to myself, "What were you thinking? Why do you think you have what it takes to be the first female operations manager, ever? These people aren't going to listen to you, they aren't going to respect you." I did it by putting one foot in front of the other, one day after another. Some days, I would think, "I've got this," and some, I would think, "What in the heck have I gotten myself into?!"

It's not enough to know your purpose and values. You must also have the courage to show up as a leader in an industry that may challenge you just because of who you are, be it young, female, black, or transgender—any number of characteristics. It takes courage to overcome challenges and make hard decisions. It takes confidence to feel that you belong and that your voice will be heard. This was one of the most common themes throughout my interviews for the book: that each of us must have the conviction that we belong in this industry and must strive for courage and confidence.

> "*Figure out what you want and don't be afraid to go after it. There are no excuses today like there might have been 20 years ago. Don't hide behind being a woman – use your skills to your advantage. Ask as many questions as you need to, without hesitation, and avoid making excuses.*"
>
> (Petrillo, 2023)

I am glad the construction industry is striving to be more accepting of diverse people and valuing these people for the diverse thinking they bring to our industry. This creates an environment that supports courage, confidence, and voice. We need to feel we belong.

Building Women Leaders: A Blueprint for Women Thriving in Construction,
First Edition. Gretchen Gagel.
© 2025 John Wiley & Sons, Inc. Published 2025 by John Wiley & Sons, Inc.

As you read this chapter, think about your sources of confidence and courage. What can you do as a leader to help others feel they belong? To ensure that people have a voice? To build the confidence of others? This is not just about each of you; it is also about how we lift other people and contribute to their success in the industry.

Belonging

In 1999, I received a phone call from Hugh Rice asking for a favor. A sports facility construction project had run into trouble, and the leaders needed someone to facilitate a meeting to reach a resolution. Hugh felt I was the right person to lead the meeting and asked that I call the owner of the construction company to firm up the details. When I called the contractor, he started asking me a series of what I call "pedigree questions"—questions that come across as condescending because they seem to imply that you do not know what you are doing. I called Hugh back and said something to the effect of, "He's not interested in my help, and frankly, I'm not interested in helping him!" Hugh said he would take care of it and asked me to facilitate the meeting, so I agreed.

It turns out the project was in deep trouble—a design/build project that should never have been design/build because of the facility's unique characteristics and because the client was not an experienced construction owner capable of making the rapid decisions necessary in a design/build model. The project was in a city with a hot construction market, and the initial estimates were obsolete, given the delays caused by slow decision-making. The subcontractors, having sniffed out money problems, were not sending their "A-teams" to the project. The contractor had just presented the client with a $20 million change order on a $40 million GMP (guaranteed maximum price) budget—yikes! The purpose of our meeting was to finalize the change order amount and determine how to complete the project. As I sat on a flight to that city, I thought, "What the heck am I going to do?" I knew I had to somehow create empathy and deal with how the team had gotten into this mess, and then quickly shift to how to get out of it.

When I arrived, I asked the CEO of the client organization and his project manager to go to one room, the CEO of the contractor and his project manager to go to another, and the principal architect to go to yet another. I gave each room three sheets of flipchart paper. The top of the first sheet said, "The client is feeling …;" the second sheet said, "The contractor is feeling …;" and the third sheet said, "The architect is feeling …" I asked each room to fill out all three sheets (one for themselves and one each for the other two parties), with a bulleted list of how they were feeling.

When we came back together, I put up all three of the flipchart sheets describing how the client was feeling, their own thoughts, and those of the contractor and architect. Incredibly, all three groups had described similar feelings. The client felt let down and disappointed, and both the contractor and the principal architect felt they had let the client down, and articulated that they knew the client was feeling let down and disappointed. I repeated the process for the contractor and principal architect with similar results. The client understood that both the contractor and architect were feeling embarrassed and that they had let their client down. We discussed how this had happened, that the contractor (in trying to be a team player) had agreed to changes without clearly communicating the cost of these changes and associated delays to the client. The contractor was late in realizing the magnitude of the cost impact of the delays—especially given how much the construction market had heated up in that city. Once we had created empathy and understanding, we could then move on to reaching an agreement on the dollar amount for the change order and how we were going to get the subcontractors back on site and motivated to complete the project.

With our tasks complete at the end of what I can only describe as a grueling 12-hour meeting, I stood in front of the group and congratulated them on the agreements they had reached. The contractor said to the group and me, "Hugh was right. You were exactly the right person to help us," to which the client's project manager quickly responded, "Tell her what you really said, that there was no way a woman could help us out of this mess," or something to that effect. The contractor admitted to saying so and apologized to me in front of the group. I give him credit for having the courage to acknowledge this to my face. More meetings ensued with the subcontractors and the project was back on track, and turned out to be a stunning athletic facility. My husband and I were invited to the black-tie opening gala, where the contractor came up to me, gave me a big hug, turned to my husband, and said, "Your wife saved this project!" It was a tremendous compliment, and I can still picture that moment.

Even when people do not believe you belong, you need to believe you belong and that you add value to your team, your organization, and the industry. Of course, people will doubt us for many reasons, including because we are women, because we are young, or because we are from another country (my experience in Australia, starting all over again at 54!). The most important person we need to believe in us is ourselves. If we do not believe in ourselves, no one else will. A male construction industry executive once told me that he believes every woman leader he meets in the construction industry is exceptional. Why? He thinks mediocre men can thrive, but to thrive as a woman, you must be exceptional. If he thinks so, we should all think so too—that we are exceptional!

> *"You need to understand that you really do belong here. It's that sense of you belong here as much as anybody else when you walk into room. I take up as much space as they do."*
>
> (Denise Burgess, speaking to Gagel, 2023 Greatness Podcast)

Many of the women I interviewed for this book told stories of men and women suggesting they raise their hand for a position when they did not consider themselves a qualified candidate, and of accepting these positions with tremendous success. It is a gift when others believe we belong, as it gives us the courage to succeed. As Maria Pilar Gomez Fabra, Human Resources Director at Acciona Australia New Zealand, shares: *"Focus on what you bring to the table and hold onto this, trust your instincts, believe in yourself. You were hired because you have something to add. Someone believes in you so believe in yourself. Build on that level of confidence and trust in yourself"* (Gomez Fabra, 2024).

The first step to courage and confidence is believing you belong and wanting to belong even when others may doubt you. Your ability to embrace that as a leader (and in turn to create an environment where others feel they belong) is powerful.

Activity:

- Write down all the things you love about the construction industry.
- Make a list of all the reasons you belong here. You might even draw upon the thoughts and ideas of other people who believe you belong here.
- Make this into a statement. "I belong in the construction industry because I love it, I'm smart, I'm a great leader, and I can help change the industry." You have your mantra!
- What actions can you take as a leader to help others feel they belong?

Courage

I remember sitting by the window at North, a Cherry Creek restaurant in Denver, with my amazing friend and mentor, Barb Grogan. I had just stepped down as assistant dean of the Daniels College of Business at the University of Denver because of my mother's health. I was feeling a bit insecure because I was single, unemployed, and had two kids about to head to college. Barb gave me some incredible advice that day. She told me I was wonderful, that many people would be beating down my door for a job when I was ready to go back to work, and that I should not settle—that I should do exactly what I wanted to do. It would all work out. And it did.

I have had many times in my life when a personal or professional decision has required courage—none more so than my move to Australia in 2018. I had absolutely no idea how I was going to earn a living in Australia. Somehow, I had faith that it was all going to work out, and it did. I believe that many of us look at other people (particularly women in our industry) and think, "Wow, she has it all together. Why am I so scared?" This was the primary reason I wrote my book, *8 Steps to Being a Great Working Mom* (Gagel, 2015). I felt that many working moms were looking at each other, thinking that everyone else had it all figured out, and that we needed more candid conversations about how hard it is to be a working parent—especially a mom.

Courage is defined as "mental or moral strength to venture, persevere, and withstand danger, fear, or difficulty" ('Courage,' 2024). We all have times when we are afraid. To persevere is to believe in yourself, your values, and your ethics, even when others are trying to tear you down. Having courage starts by recognizing your skills and the value you bring to your team and organization. It begins by acknowledging your fears, that we all have them, and that you are not alone in these feelings. It starts with taking that one step that transforms fear into confidence.

> *"For the Denver International Airport Westin Hotel, we received the largest contract for $39 million. The Denver Post called and asked me a few questions and the next day there's this front-page Sunday Post above the fold article stating, 'She does not know what she's doing, she's not capable, she's not capable of managing firms, she has no ability'. Now keep in mind that I had already successfully completed two contracts of that size in different parts of the country, on time and on budget. Suddenly investigations were launched, and I was audited 1,200 times by the City Auditor during the course of that project. To have that assault on you personally, first you have to get up off the bathroom floor because they just have knocked you to your knees. To have someone say we're going to do an internal investigation on your firm and then to have to go to your employees and your managers who now have to go on site and execute this project and they were terrified. I had to first pull them together and say, 'Let's just do what we do. We know how to do it.' Then to have your family read it. You quickly learn who are friends and not friends in the circumstances, and you quickly learn your resolve to say, 'You know what? All I have is what I know how to do.' You just have to keep that in mind, that you're a hundred percent qualified and that's what I had to do every day. We actually ended up training three minority and women-owned businesses that had their first project with the Denver International Airport, and they're still there today working, so I'm very proud of that. But I must tell you it's those firestorms that will teach you your worth. And they'll also teach you the value of what your parents have taught you which is to believe in yourself 100%."*
>
> (Denise Burgess, speaking to Gagel, 2023 Greatness Podcast)

This is courage, to face a scathing front-page article and persevere. Hopefully, most of the challenges you and I face will not be the "firestorm" that Denise was up against—but if she can persevere, so can we. It starts with knowing your craft. *"The technical acumen I learned early was important so that I was prepared for anything. It's assumed that we don't know as much about construction. I encourage women to put your head down and learn your craft. You don't have to prove you're the smartest person in the room, you just have to be ready to answer any question asked of you, or offer a solution for consideration"* (Neuscheler, 2023). As friend and Founder of Monarch Build, Courtney Kounkel, shares when asked for her advice to women in the construction industry: *"Ask a lot of questions. Don't be afraid of asking questions because everyone's learning. Be confident in the things you do know. Don't be quiet and sit in the back of the room. Be a woman, it's your superpower. You're not better, it's what's unique about you, what you bring. Be excited"* (Kounkel, 2023).

"I had a pretty significant fall in 2008 leading into the Beijing Olympic Games. I fell at high speed in the keirin in Los Angeles in a World Cup final, and to top of the list of injuries that came with that fall was a broken neck. I fractured my C2 vertebra which is the second down from the skull, and I learnt some weeks after the fall that I was two millimetres from a clean break. And the subsequent recovery and rehabilitation wasn't just a physical one, it was a mental one and an emotional one as a result of the enormity of those injuries. I only learnt that because I overheard a conversation between my coach and the medical team at the time and the doctor actually said to my coach, his exact words, 'God has other plans for this young lady because she was two millimetres from a clean break', which had that two millimetres not have been there, I would have been at best a quadriplegic requiring a respirator for the rest of my life to breathe and the only other option was death.

When I started to really comprehend the severity of that as a young adult, it really smacked me, and I realized I was very struck by fear, and I started to think in the context of 'what if'. What if those two millimetres hadn't been there? What would my life have been like? Would I have been happy with my life up until that point? And this is how I started to engage in discussion with my coach after I overheard this conversation. He was very calm, and he just let me talk to him first before he gave any input, and all he said was, 'Anna, you're asking the right question, you're just using one wrong word. Don't ask yourself "what if," ask yourself "what is".' He went on to explain the difference in the two. 'What if' is great, especially in business context for contingency planning. But when you're looking at situational responses to certain circumstances and particular moments that happen in time, which this was one of them, 'What if' is actually driven by fear and doubt of what we don't want to have happen. 'What if that two

> *millimetres hadn't been there? 'What if the play puts me in front of the goal, and I miss the goal for the team? 'What if I go for the job interview and I don't get the job interview? These are the sort of scenarios that come into 'What if' context and all of those things haven't happened yet. When we focus on the negative outcome driven by fear and doubt of the question 'what if', often we will have an emotional, physical response which will steer our actions and choices down the wrong path.*
>
> *Contrarily if you think situationally in the context of 'what is', you can only deal with the reality or the real tangible information that you have to deal with at the time. The 'what is' of my situation was simply that two millimetres saved my life, and when he was able to put that to me it was a real eye-opening moment. Firstly, how powerful our minds can be when driven by emotion and fear and doubt. But secondly, how your outlook to a situation can completely change the path and direction you end up taking. So, I did, I went from being struck by fear and doubt to thinking, yeah, from the 'what if' to the 'what is' of reality of that situation and I followed the 'what is', the tough path of that response, and had I not done that I would not have gotten through the rehabilitation and got to those Olympic Games seven months later."*
>
> (Anna Meares, speaking to Gagel, 2020 Greatness Podcast)

We control the voice in our head, how we speak to ourselves, the words we use, and how these words contribute to either fear or courage. We all have fears—like snakes or spiders—and that is okay. It is okay as a leader to feel uncertain, to feel anxious about a decision, and to dread a difficult discussion. These are all valid feelings and emotions. Great leaders frame emotions in a way that fuels their performance and are adept at dealing with fear and having the courage to triumph.

Activity:

- Think about a time when either yourself or someone you work with has had the courage to persevere during a challenge. What resources (colleagues, faith, etc.) did you or they rely upon?
- What do you believe are your sources of courage?
- How can you go about congratulating yourself for having courage?

Confidence with Humility

I am anonymously evaluated after every course I teach at the University of Denver and the Australian National University. One year, a student stated in the comments that he thought I was "arrogant." At first, I was taken aback. Arrogant, how dare he! And then I was hurt. I can't possibly be arrogant. I pride myself on humility. And then I thought about it and realized, yes, to be successful in the construction industry, I have had to be a highly confident woman. I walk a fine line between confidence and arrogance, trying hard to stay on the confidence side, but I sometimes probably do stray over to what some may view as "arrogant." The lesson here is that every piece of feedback has at least a grain of truth that should be reflected upon, and one person's opinion does not constitute reality. I am okay being a highly confident woman who is occasionally viewed as arrogant.

I have often wondered about the source of my confidence, and can honestly say I have no idea. I can also share with full transparency that even I, someone people describe as "highly confident," have many moments of self-doubt. I have had many of them while writing this book! Your ability to build your confidence (and the confidence of those around you) is a critical component of effective leadership.

> "*I think it's very important for females to be confident and assertive and to self-promote. I think that has really helped me. It takes time to be confident and to embrace yourself, especially as a young professional. I remember when I joined the industry, I was very quiet and very reserved. But as soon as I found my voice I used it and I think it's really important for them to use their voice when they find it, and to be authentic, and to always support each other.*"
>
> (Laura Miranda, speaking to Gagel, 2024 Greatness Podcast)

Confidence is defined as "a feeling or consciousness of one's powers or of reliance on one's circumstances," "faith or belief that one will act in a right, proper, or effective way," and "the quality or state of being certain" ('Confidence,' 2024). Research indicates that genetics may play a big part in how confident we are, and that other factors such as our upbringing and environment contribute to our confidence as well (ROAR!, 2020). Confidence levels may also change in different circumstances. My confidence dropped each time I changed industries, and I thought to myself, "Will I know what I'm doing? Will people accept me?" This ebb and flow of confidence is natural. Confidence can also improve. During one of our participant surveys of the APGA Women's Leadership and Development Program, 90% of the participants reported an increase in confidence.

> Friend, author, and speaker Michelle Sales shares this on confidence: "*It's not enough to just be the quiet achiever. You have to build confidence in a grounded way to be able to show up and have the right level of influence and impact.*

> *There are thousands of definitions of confidence and I think it's easier in some ways to start with what it's not and it's really important to distinguish between confidence and arrogance. For me, confidence is the ability to be your best self, to know what you stand for and what's core to you, to be able to have a voice that has an impact, and to be able to perform at our best. I have this belief that competence and confidence sit side by side and they're so important to us being able to do what we're born to do. My research points to people having the right balance of confidence as able to show up in a way that is genuine and grounded and connected to people, and able to exercise leadership with that confidence but engage people with elements of empathy and compassion and humility."*
>
> (Michelle Sales, speaking to Gagel, 2020 Greatness Podcast; Sales, 2022)

Here are some thoughts on how to improve our confidence and the confidence of others:

- Gaining Credentials: During my interviews with women for this book, one tactic for increasing confidence was having credentials such as degrees and certificates that help prove our knowledge to others. Often, these credentials are more for us than others to generate feelings of confidence. As Kelly Blackwell, Director, Strategic Sourcing & Procurement - Facilities, Real Estate and EOHSS at Bristol Myers Squibb shares: *"Lead with confidence. It is still a male-dominated industry, but be confident in your education, your training, your knowledge"* (Blackwell, 2023).
- Continuous Learning: Our ability to learn on the job—to prove to others that we are serious about our craft and our industry—is another important strategy for building confidence. *"I created my own opportunities to learn, sat in on job site meetings, did other work on my own time. That's how you get exposed to things and it helps to think, 'I'm going to learn more today'. That was a choice for me. No one's path is the same, we all get there a different way. You have to do it your way and be authentic to yourself"* (Neuscheler, 2023).
- Being Prepared: Being prepared is another strategy for building confidence. I am confident going into speeches and meetings when I know I have done the preparation. I facilitated a Construction Industry Institute (CII) conference CEO panel featuring Steve Edwards, then-CEO of Black & Veatch; Greg Bentley, Founder and CEO of Bentley; and the head of the construction industry sector for McKinsey. I prepared by interviewing each of the panelists ahead of time and was confident that my questions would generate a lively discussion. What I didn't prepare for was being onstage in a skirt with high-top chairs and no tables. The things men do not need to think about!

> As Sharmeena A. Salam-Haughton, Branch Chief, Construction Management, US Department of State Bureau of Overseas Buildings Operations, shares: *"Be confident in your abilities and communicate that confidence. In the beginning I tended to take a back seat, let other people communicate their ideas. Even if you're wrong, you've communicated your ideas."*
>
> (Salam-Haughton, 2023)

Preparation also provides the confidence to make bigger life changes. Before I left FMI to become the President of The Women's Foundation of Colorado (WFCO), I spent two years in night school earning a Masters of Nonprofit Management. Not only did this help prepare me for the job and give me the confidence I needed to make the switch, it also allowed me to determine that what I really cared about in the nonprofit sector was women and education, and led me to a job that matched my passions.

- Facing Fears: Facing our fears, accomplishing tasks when there is fear associated with the task, also builds confidence. *"I share the story in my book The Leading Edge about spending a whole year, three hundred and sixty-five days, doing something I was afraid of and the unbelievable learning journey that took me on and the lessons that I learned around the need we all have to re-sensitize to fear. It's a word that's sort of been dramatized by Hollywood. We think about jumping out of planes and spiders and snakes but let's be real, that's not the barrier that's getting between any of us and our goals"* (Holly Ransom, speaking to Gagel, 2022 Greatness Podcast).
- Building a Support Network: The support of others is a huge source of confidence—those raving fans who say, "You've got this." We will discuss this in more detail later in the book.

We can also physically do things that convey more confidence.

- One important strategy people (especially women) can use to come across as more confident is speaking a little bit more loudly. That does not mean you have to be the loudest person in the room—just turn up the volume. I was recently sitting in a meeting during which a woman was leading us through a difficult conversation. I thought to myself, "She's not speaking loudly enough!" I could barely hear her, and to command confidence from those of us sitting around the table, she needed to speak up, just a bit. As Micki Kohn, Executive Vice President of Project and Support Services for Hargrove Engineers and Constructors, shares: *"Don't be bashful about speaking up as it's important to represent yourself. Respect others and expect them to treat you with respect. Don't let it roll off your back – inappropriate comments. To me it's important that it's understood that it is not okay, that it's clear that you expect to be treated with respect"* (Kohn, 2023).

Presentation coach and author Joel Schwartzberg shares an exercise he uses to demonstrate how volume alone can increase perceptions of confidence: *"There is a part in my workshops where I encourage people to speak louder. The reason I encourage everybody to speak louder – especially for people who don't normally speak loudly – is because amazing things happen to your impression. When I ask everyone else in the room how that person's impression changed, they say the person seemed more confident, more assertive, more authoritative, more credible, and more like a leader. Someone who's a coordinator suddenly sounds like a senior manager. And to be clear, the person who raises their volume may not be more confident, may not even feel more assertive, and may not even feel like a leader, but just by merely going louder, they create that impression. I've done this test for over 15 years and every time I poll people on how volume increased someone's impression, they say increased assertiveness, power, presence, competence, and authority.*

Sometimes there's a gender issue. Some women in my workshops have said they fear that being louder will be perceived as aggressive, which is a negative. My response is that all people should speak up and use their volume to come across with power and confidence. Even if someone in your audience has the sexist attitude that a louder woman sounds aggressive, that's their problem, not your problem. You always need to speak up, even if those people need to grow up."

(Joel Schwartzberg, speaking to Gagel, 2023 Greatness Podcast)

- Body language can also convey confidence. Many of you have probably seen Amy Cuddy's (2020) TEDx Talk on power stances, and they work! A few years ago, we were conducting a study in the building materials industry that entailed cold-calling companies to gather information. One of my colleagues was feeling a bit intimidated by the process. She started doing "power stances" before each call, and it truly boosted her confidence and the results she achieved. Just by leaning forward during a meeting, we can convey more confidence. Are you sitting tall or slouching? Does your body language convey strength or weakness?

I keep a file folder (physical and electronic) of all the nice things people say to me. It comes in handy some days when you need a boost of confidence!

One of the questions I hear most frequently from women—particularly in Australia, where "tall poppy syndrome" describes a geographic culture that at times does not value one person rising too high above the others—is, "How can I be confident without coming across as arrogant?" Arrogance is defined as "an attitude of superiority manifested in an overbearing manner or in presumptuous claims or assumptions" ('Arrogance,' 2024). Yikes—overbearing and presumptuous? Superiority? None of us want that!

As women, I believe we have experienced situations where our behavior is condemned as being "arrogant" or "too assertive," when that same behavior in someone who identifies as a man would be condoned (if not applauded), and would certainly not be viewed as arrogant. In 1988, I was 23 years old and running a cereal plant, and I was reporting to one of the more senior operations managers who managed another plant on site. During one of my performance reviews, this man told me I was too aggressive, too assertive. I pointed out to him that he liked to pound his fist on the table and yell @#$%. His response? He could do that because he was a man, but if I was too assertive, I was just being a ^@#$%! Wow. Yes, it's good to worry about not being perceived as too arrogant, too aggressive—but not to the detriment of your confidence.

PhD sociologist, author, and thought leader Dr. Tracy Brower shared her research on arrogance and "likeability bias" during a visit to the Greatness Podcast: "*What led me down the path was some really interesting research I came across that looked at when leaders are arrogant, things like retention and performance suffer; and some additional research on different types of arrogance. Individual arrogance is more about when you just feel an over-inflated sense of your personal value; comparative arrogance is when you feel better than others; or antagonistic arrogance is where you want to cut other people down.*

One of the challenges for women is there is a likeability bias. There's some fascinating research that suggests that if people dislike women, they are more likely to not cooperate, to not bring their best, to not want to be part of the project or contribute to the positive ends of the project. The same is not true for men. If men are unlikeable, it doesn't necessarily cause a reduction in the cooperation that men see around them.

"So how do women face this challenge of being confident, and having great ideas they want to bring forward without being perceived negatively? I think one of the differences between confidence and arrogance is how other people are receiving the confidence that we bring across. Confidence is all about, 'I believe in my ideas and I'm going to assert them with a level of confidence', without, 'I'm not assuming that my ideas are better than yours' arrogance. Arrogance is, 'I've got the right idea and everybody else may not have as good an idea', or 'I may not have something to learn from other people'. There's an element of humility that is inherently part of that. 'I think I've got a great idea and I'm confident about my performance but I'm curious about what you're bringing', 'I'm assuming that there's

> *something out there that I don't already know', 'I have humility to learn from others as well as being confident in my own point of view'. I think that humility aspect is an ingredient as we look at an absence of arrogance."*
>
> (Tracy Brower, speaking to Gagel, 2024 Greatness Podcast)

Research supports humility as an important ingredient in great leadership. Amy Edmondson, in her groundbreaking research on psychological safety and the impact of it on team performance, identified situational humility as a critical leadership trait (Edmondson, 2018). It is okay to say, "I don't have the answer." No one has all the answers, and faking it is disingenuous and negatively impacts trust. Jim Collins (2001) describes a "Level 5 Leader" as a leader who "displays a powerful mixture of personal humility and indomitable will." The key is to know when to use humility, when to use confidence, and to what level for each. To be humble is to be vulnerable, which is another key attribute of great leaders. In fact, according to research, leaders who show vulnerability cause their employees to be 5.3 times more likely to trust them (Fourmy, 2023).

Balancing confidence and humility can be a challenge—especially for women in male-dominated industries. I encourage you to embrace the confidence you have while maintaining the ability to be vulnerable. Humility is critical.

Activity:

- On a scale of 1 to 10, with 1 being "low confidence" and 10 being "high confidence," what is your general confidence level?
- When do you feel most confident? Why is that? How can you parlay that "why" into greater confidence during times of low confidence?
- What next step might you take to become more confident?
- How do you go about demonstrating humility? In what ways might you improve?

A Note on Imposter Syndrome

Research supports that 70% of us will experience imposter syndrome in our lives—a deep-rooted feeling that you are an intellectual fraud or fake (despite evidence of competence) and that you will be "found out" as someone who does not know what they are doing (Alison Shamir, speaking to Gagel, 2022 Greatness Podcast).

> *"Don't be afraid to ask for opportunities. You are in charge of your career. My boss's boss told me 'Don't only get to know your boss's boss, get to know your bosses' bosses boss.' Be in the succession plan and don't be afraid to ask for development opportunities. Stop apologizing! We over-apologize. Ask for a project, a position, development. Be more confident, find the tools. I had a horrible bout of imposter syndrome during my career but didn't let lack of confidence or imposter syndrome stop me."*
>
> (Hinz, 2023)

My friend and global imposter syndrome expert Alison Shamir helped me understand the difference between lack of confidence and imposter syndrome (Alison Shamir, speaking to Gagel, 2022 Greatness Podcast). Alison is incredibly authentic in her work because it was her own severe imposter syndrome that led to a mental breakdown. Her curiosity about how to heal herself led to her research and life's work—and that is the good news: we can overcome imposter syndrome. Alison shared her thoughts on overcoming imposter syndrome during a visit to the Greatness Podcast:

"When you experience imposter syndrome you feel like you're fooling everybody and there is this underlying fear of being exposed or found out to be an intellectual fake, phony, or fraud, and to not be good enough or worthy enough of your success. The irony is those of us who experience it are competent. We believe we're not worthy because something has happened in our past to make us feel less than, so the accomplishments and the success and the accolades or whatever is coming our way don't quash that belief system. We actually have to do the work on that belief system. We catastrophize how we view failure. We try to be a perfectionist because we use that as a shield. If we're perfect and if we strive for perfection, no one will ever know that we're a fraud.

You can absolutely move through it and conquer it. You need to find somebody that you trust, find somebody who offers you a psychologically safe space. This could be one person, could be a group of people, could be the women's leadership program that you're in, and share your journey because somebody around you is having a similar journey. Getting out of your own head and getting into the practical sense of vocalizing and communicating and essentially problem solving your way through imposter syndrome is powerful. We can't battle it in our heads because it already takes over our thoughts which impacts our behaviors. We've got to get out of there into practical action. Normalize the conversation, elevate the conversation, find someone that you trust.

The other thing that I recommend is to have a look at what's triggering your imposter feelings. We can develop strategies to help you create boundaries or to help you manage the way that you maneuver around that trigger before you self-sabotage.

We need to intercept the negative behavior and you can do that mainly by leaning back on a support network or sitting down with yourself if you believe you can, and often we can brainstorm on our own and saying, 'okay rather than doing that negative behavior, what is one thing, one micro thing that I can do differently that might break that cycle?' We get to rewrite our story and we can then rewrite the belief system. If we believe we're not worthy, we're not just going to wake up tomorrow and suddenly go, 'I am worthy today'. There's going to be some steps in between. Having imposter syndrome, full stop, is not your fault. It's a product of something that happened in your past that is now coming out of you because of something that has triggered you in your present. It is not your fault" (Alison Shamir, speaking to Gagel, 2022 Greatness Podcast).

Conclusion

These topics—belonging, courage, and confidence—are critical. I believe that each of us plays a role in creating an inclusive industry where everyone feels they belong. Each of us has the inner strength and support network to be courageous, confident members of the construction industry and to have voice and thrive. Each of us supporting one another—all people in the construction industry—is a gift.

> As Meg Redwin, Executive Director and General Counsel, Global, at Multiplex, shares: *"My advice to women in the construction industry is don't have regrets. Don't think, 'Why didn't I say that?' Have the courage to speak up, to put your hand up for a role. And don't be too hard on yourself."*
>
> (Redwin, 2023)

References

'Arrogance.' (2024). *Merriam-Webster Online Dictionary*. Available at: https://www.merriam-webster.com/dictionary/arrogance [Accessed: 29 May 2024].

Blackwell, K. (2023). Interview by Gretchen Gagel [Zoom], 6 December.

Collins, J. (2001). *Good to great: Why some companies make the leap...and others don't*. New York: Harper Business.

'Confidence.' (2024). *Merriam-Webster Online Dictionary*. Available at: https://www.merriam-webster.com/dictionary/confidence [Accessed: 29 May 2024].

'Courage.' (2024). *Merriam-Webster Online Dictionary*. Available at: https://www.merriam-webster.com/dictionary/courage [Accessed: 29 May 2024].

Cuddy, A. (2020). *Your body language may shape who you are.* Available at: https://www.ted.com/talks/amy_cuddy_your_body_language_may_shape_who_you_are?language=en [Accessed: 29 May 2024].

Edmondson, A. C. (2018). *The fearless organization: Creating psychological safety in the workplace for learning, innovation, and growth.* Hoboken, NJ: John Wiley & Sons.

Fourmy, R. (2023).'Why executives need to practice vulnerable leadership—and how to do it,' *DDI Blog*, 24 August. Available at: https://www.ddiworld.com/blog/vulnerable-leadership [Accessed: 29 May 2024].

Gagel, G. (2015). *8 steps to being a great working mom.* Atlanta: BDI Publishers.

Gagel, G. and Brower, T. (2024). *Dr. Tracy Brower discusses confidence versus arrogance and her book The Secrets to Happiness at Work* [Podcast]. 31 May. Available at: https://open.spotify.com/episode/7lSFYWTciCKlTNfsdcF2Uu [Accessed: 31 May 2024].

Gagel, G. and Burgess, D. (2023). *Denise Burgess discusses her career as an African American woman leader in the construction industry* [Podcast]. 29 December. Available at: https://open.spotify.com/episode/77UCAi1FXUKeMOS6yfes8z [Accessed: 2 May 2024].

Gagel, G. and Meares, A. (2020). *Anna Meares discusses leadership and team lessons from being an Olympic athlete* [Podcast]. 17 February. Available at: https://open.spotify.com/episode/3k1RqB8K3aiLVG5lRgkRol [Accessed: 28 April 2024].

Gagel, G. and Miranda, L. (2024). *Laura Miranda discusses her podcast "You Don't Look Like an Engineer"* [Podcast]. 8 March. Available at: https://open.spotify.com/episode/6sLfjJDdlYzWoT5zkSD1nR [Accessed: 2 May 2024].

Gagel, G. and Ransom, H. (2022). *Holly Ransom discusses her book The Leading Edge* [Podcast]. 30 September. Available at: https://open.spotify.com/episode/0M5G6yRwqvbBlKa302wmKb [Accessed: 1 May 2024].

Gagel, G. and Sales, M. (2020). *Michelle Sales discusses her book The Power of Real Confidence* [Podcast]. 11 December. Available at: https://open.spotify.com/episode/79XJIyGSMH9TVysaGJjA8V [Accessed: 28 April 2024].

Gagel, G. and Schwartzberg, J. (2023). *Joel Schwartzberg discusses his book Get to the Point* [Podcast]. 16 June. Available at: https://open.spotify.com/episode/08Zpsyp0J1PZ4sKvU60w58 [Accessed: 2 May 2024].

Gagel, G. and Shamir, A. (2022). *Alison Shamir discusses imposter syndrome* [Podcast]. 5 August. Available at: https://open.spotify.com/episode/2wEPGFTmk3EN0WXLa3mml7 [Accessed: 1 May 2024].

Gomez Fabra, M. P. (2024). Interview by Gretchen Gagel [Zoom], 24 January.

Hinz, L. (2023). Interview by Gretchen Gagel [Zoom], 6 December.

Kohn, M. (2023). Interview by Gretchen Gagel [Zoom], 24 October.

Kounkel, C. (2023). Interview by Gretchen Gagel [Zoom], 7 December.

Neuscheler, K. (2023). Interview by Gretchen Gagel [Zoom], 13 December.

Petrillo, K. S. (2023). Interview by Gretchen Gagel [Zoom], 20 November.

Redwin, M. (2023). Interview by Gretchen Gagel [Zoom], 1 December.

ROAR! (2020). *Are some people just born confident?* Available at: https://roar.training/are-some-people-just-born-confident/ [Accessed: 29 May 2024].

Salam-Haughton, S. A. (2023). Interview by Gretchen Gagel [Zoom], 26 October.

Sales, M. (2022). *The power of real confidence: Learn how to lead to your full potential.* Melbourne: Major Street Publishing.

5

Thinking Critically, Learning, and Reflecting

Introduction

I was working with the leadership team of a multi-billion-dollar construction program where speed of execution to increase critical manufacturing capacity was pivotal. One day, the executive leadership team was informed that a large portion of one of the construction sites' foundation had been poured incorrectly and would need to be torn out and replaced. Emotions could have reigned, but instead, I witnessed leaders who remained calm and employed critical thinking skills to focus on the facts of the situation, leveraged their problem-solving skills to move quickly to a solution, and then took responsibility for learning as leaders. This is a wonderful example of leaders utilizing their brains to think critically, learn, and reflect.

My friend, author and speaker Kristen Hansen—the only person I know who takes a plastic brain with her everywhere—has helped me understand how research and functional magnetic resonance imaging (fMRIs) are unveiling the inner workings of our brains. I believe the next frontier of leadership knowledge lies in neuroscience, and that leaders must seek to understand and fully utilize our brains to think critically, make decisions, learn, gather feedback, learn more, and reflect. Your development as a leader is a lifelong journey, and leveraging your brain on that journey is essential.

> "*I like to think that if you understand the mechanics of how the brain works, not only your own but other people's brains, you're in a better position to be able to make adjustments to get the best out of yourself and the best out of others. I think leaders really need to show up in a very positive way. They need to be extremely resilient and agile. If we can understand what makes the brain become more agile,*

> *what makes the brain become more resilient, what helps us have better conversations by ensuring that somebody is in a receptive mode to new information or to creativity etcetera; if we can actually unpack that and understand all of these mechanics behind great performance then we're in a position to be able to make adjustments for ourself and make adjustments for the way we communicate and create environments for others to succeed."*
>
> (Kristen Hansen, speaking to Gagel, 2019 Greatness Podcast)

I have divided this chapter into two sections ("Utilizing Your Brain" and "Caring for Your Brain"), as I believe both are critical. As you read through this chapter, consider your strengths and opportunities related to how effective you are at leveraging critical thinking. Additionally, think about your adeptness at resting your brain to ensure you are on your game mentally and fueling creativity.

Utilizing Your Brain – Critical Thinking

Critical thinking is by far the most important skill I learned during my PhD. Construction projects have multitudes of subcontractors who are often arguing over resources such as lay-down areas or providing seemingly-conflicting information on the constructability of a design, or simply sharing differing opinions on how we should construct the project. We are constantly bombarded with opinions, half-facts, assumptions, and emotions. Our ability to sort through the white noise and discern what information is important, correct, and helpful is vital to our ability to be great leaders. We ingest about 100,000 words a day, and multitudes of thoughts and ideas—some of it "propaganda" as societal influencers work to sway our opinions on anything from what shampoo we should use to whether global warming is a thing or not (Ramsey, 2009).

Critical thinking is defined as "…the act or practice of thinking critically (as by applying reason and questioning assumptions) in order to solve problems, evaluate information, discern biases, etc." ('Critical Thinking,' 2024). There are a couple of important points in this definition. First, critical thinking is a "practice of thinking," and it does take practice ('Critical Thinking,' 2024). Unfortunately, the US education system often does not encourage critical thinking. When I was taking design of steel structures in engineering school, we were handed a set of equations to determine how large an "H" or "I" beam to use, given a certain set of parameters for the building we were designing. You did not question the equations; you just used them. A second point in this definition, to "apply reason and question assumptions," is a simple statement that can be challenging to enact

('Critical Thinking,' 2024). Questioning assumptions goes back to our discussion of unconscious bias. We all make assumptions and have underlying biases that color our thinking on topics. The key is our ability to examine these assumptions and keep assumptions from narrowing our thinking and decision-making.

> When using critical thinking, it is important to separate facts from opinions. As a professor, I would continually challenge my students: when you write a sentence in a paper (such as "leaders are good listeners"), the statement is either your opinion and needs to be written as "I believe that leaders are good listeners," or is stated as a fact with a reference or evidence to support the claim like, "Research by XYZ person supports the idea that leaders are good listeners." This concept sounds simple, but think about the number of items you read that are stated as fact, but are actually unsubstantiated opinions. *You've got to nurture and protect the inner integrity of your intellect, your ability to think, to formulate ideas and then express them in a way that can challenge orthodoxy. This is critically important."*
>
> (Brendan Nelson, speaking to Gagel, 2019 Greatness Podcast)

Let us think about critical thinking as it applies to a construction project. Someone comes to you and says, "Charlie isn't being safe on the jobsite." That is an opinion. The facts might be that Charlie's safety helmet chinstrap is undone, or Charlie is not wearing safety shoes. These are the facts of the situation and can be addressed. Someone might come to you and say, "We aren't going to meet our deadline for temporary occupancy for Building 3." That is someone's opinion, better stated as, "I don't think we are going to meet our deadline for temporary occupancy for Building 3." The next step is to dig into the facts of the situation rather than "go to drama," to have an emotional reaction to an opinion rather than calmly uncovering the facts of a situation. Critical thinking demands that we make fact-based decisions, and teams appreciate this in their leaders.

Several frameworks of critical thinking exist. I typically use Bloom's Taxonomy of Critical Thinking, created by Benjamin Bloom, an American educational psychologist ('What is Bloom's Taxonomy?', n.d.). Here is a brief overview:

- **Remember:** The first of three "tier one" levels of thinking is to remember a piece of knowledge, like remembering the formula I used to calculate the size of a steel beam.
- **Understand:** It is not enough to remember this piece of knowledge; you must understand it, like what the different letters and symbols mean in the formula.
- **Apply:** Next, you apply the knowledge, like applying the formula to calculate the correct size beam.

- **Analyze:** The first of three "tier two" levels of thinking brings us to the level of critical thinking. This involves breaking a thought, idea, or piece of information down to its fundamental elements in order to understand it. For example, we might examine each element of the formula.
- **Evaluate:** As we evaluate the information, we start to think, "Do I agree with this?" or, "Is this information supported by evidence or other pieces of information I have stored in my head?" We use evidence to evaluate an idea—not opinions or assumptions.
- **Create:** At this level of critical thinking, we are creating new knowledge and ideas through our analysis and evaluation of existing thinking. We might, for instance, create a new formula that is more effective in the design of steel structures.

Our ability to operate at the tier two level of critical thinking—to go beyond understanding and applying information to the level of analyzing, evaluating, and creating knowledge—is a critical competency.

> One day, I was sitting on a flight and the NFL player sitting next to me took out a set of index cards. One side of each card contained words that this player reviewed, and I could tell he was memorizing the words to **"remember"** the play. He would then flip the index card over, and a play was drawn on the back in X's and O's to help him **"understand"** the play. The next logical step would be to **"apply"** that knowledge by accurately executing the play during a game. This player's thinking could stop there at "tier one" critical thinking. He could be a great player who remembers the play and effectively executes it during games.
>
> But what if during a game, the play isn't working? What then? Maybe that player needs to **"analyze"** the elements of the play and **"evaluate"** why it isn't working? Football players frequently appear on the sidelines looking at rugged tablets to evaluate plays, and coaches are adept at using statistics to evaluate what is working and what is not. This is why we have halftime! What if this player took his thinking a step beyond "tier one" thinking, approached his coach, and suggested a change? He might **"create"** a new play that is a combination of two other plays. That is "tier two" critical thinking. Who do you want on your team—the player who is excellent at execution but does not think beyond the tasks that are given to them, or the player who uses their brain to help create new knowledge, new ways of executing work that are more effective?

You as a leader set the example for your entire team via your ability to think critically by sticking to the facts and understanding your assumptions (while being emotionally intelligent!). You encourage people to think beyond today's known knowledge to analyze, evaluate, and create new knowledge. Your ability to think critically is one key element of great leadership. As with everything else, it requires that you be mindful of what critical thinking is and develop it as a skill.

Activity:

- Take a few minutes to think about what level of Bloom's Taxonomy of Critical Thinking you typically operate within.
- Are you adept at sorting opinion from fact? Are you evaluating the information you ingest to formulate your own opinions? Are you present to your assumptions? Your bias?
- What is one step you might take to improve your critical thinking skills?

Utilizing Your Brain – Learning and Curiosity

Three of the people I most admire in the construction industry are the late Peter Nosler, co-founder of DPR Construction; Jan Tuchman, retired Editor-in-Chief of *Engineering News-Record* (ENR); and Hugh Rice, Chair Emeritus of FMI Inc. I am grateful for the many meals I have had with each of these individuals who exemplify what it means to be a learning leader. Great leaders continue to learn and be curious about the world. Great leaders remain open to what they do not know instead of focusing entirely on what they do know—a concept described by Carol Dweck as a "growth mindset" versus a "fixed mindset" (Dweck, 2007).

> *"In a fixed mindset, people believe their basic qualities, like their intelligence or talent, are simply fixed traits. They spend their time documenting their intelligence or talent instead of developing them. They also believe that talent alone creates success—without effort."*
>
> (Dweck, 2007)
>
> *"In a growth mindset, people believe that their most basic abilities can be developed through dedication and hard work—brains and talent are just the starting point. This view creates a love of learning and a resilience that is essential for great accomplishment."*
>
> (Dweck, 2007)

Great leaders consider both formal and informal learning to be lifelong journeys. As a first-generation college graduate, I value the opportunities my formal education has afforded me. I set a goal decades ago to earn a college degree every 10 years, and after four university degrees, I think I'm done! Each of my college degrees had a specific purpose and learning object. My engineering degree satisfied my desire to study the sciences and prove my father wrong—girls do engineering! Six years later, I was managing a $28 million profit and loss statement without any formal business training, thus the MBA with a focus on finance. My Master's in Nonprofit Management facilitated my move into the nonprofit sector, and my PhD in leadership, organization culture, and agility/change management came about because I wanted to study the research on how leaders create nimble organizations that can morph and thrive in shifting business environments. Each degree had a purpose and expanded my knowledge, and was driven by my continued curiosity, which I believe also contributes to great leadership.

> *"Grandma has often described me as forensically curious, which I think is pretty apt. But I didn't grow up with a dynamic that meant those questions were welcome around the dinner table and so I was always someone that was seeking answers outwardly. I was trying to find out, what can my teacher tell me? What can my sports coach teach me? What is it that the person who works at the local shop can teach me? I was always hunting for answers but outwardly because that had to become my learning playground, and I think in a lot of ways that's been probably one of the most helpful muscles that in many ways I didn't realize I was building. It was just sort of a desire for answers that led to it originally. But that idea that everyone can be a source of information, that idea that we need to actively seek mentors and learning in a diverse manner, that in many ways that has shaped my approach to life, has brought me in contact with some remarkable individuals, and has made me incredibly passionate about democratising access to that information because it's transformed my life and I've seen the power of it transforming others too."*
>
> (Holly Ransom, speaking to Gagel, 2022 Greatness Podcast)

When you model the behavior of curiosity as a leader, you set the tone for organizational learning and curiosity. This is a key factor in an organization's ability to remain relevant and achieve sustained success because I believe organizations that employ people who remain curious and continue to learn are more prepared to survive shifts in the business environment. It starts with the leaders. As friend, author, and speaker Dr. Pamela Meyer shares: *"The easiest way to understand agile leadership is to think of the opposite of it. The disembodied leader walks into the room with this mindset of 'I hope nothing goes wrong', whereas the agile leader walks into the room thinking 'what's going to happen today? What am I going to learn? What are we going to create today?' They're open to surprise and possibility, and that is very much a different mindset. It's the learning and adaptation mindset versus a planning and control mindset and honestly the two aren't mutually exclusive"* (Pamela Meyer, speaking to Gagel, 2023 Greatness Podcast).

I believe that informal, less structured learning can be just as effective and fulfilling as you develop as a leader. In the last five years, I have taken online courses in Qigong, meditation, hypnotherapy, and botanical watercolors. The last is a nod to my opinion that I believe we all need to have a creative outlet of some kind. Here are some ideas on how you can remain curious and continue your informal learning as a leader:

- Spend time with an employee learning a new skill (or take a class in this skill). When I was managing the baby food plant, I took a course in "retorting" to learn how we cooked baby food.
- Set a goal as a team to learn something new. We did this at Continuum Advisory Group when we tackled emotional intelligence as a team. Team learning is fun, upskills the entire team, and provides you with a chance to practice together.
- Learn a new technology. My good friend and author Donna McGeorge was asked by her publisher to learn about and write a book on ChatGPT, and she did (McGeorge, 2023). In the construction industry, there are ample opportunities to learn about new technologies.
- Attend a conference that is not related to your industry. I remember one of my management consulting team members attending the Salesforce conference one year. Beyond learning about Salesforce (our customer relationship management software [CRM] system at the time), this person returned brimming with ideas on a variety of other topics.
- Use an app to learn a new language. This might come in handy both on vacation and at work. For example, if you are involved in construction in Colorado, knowing Spanish is advantageous.
- Take an online course in meditation, Tai Chi, Qigong, or breathing exercises. These techniques are fun, and also help us cope with stress and improve our cognitive ability and emotional intelligence.
- Learn how to play an instrument, paint, or start some other creative outlet. There is quite a bit of research supporting how these creative outlets support our ability to be great leaders (Baskin, 2022).

Our ability to remain curious and willing to learning positions us for continued success as leaders. Maintaining a growth mindset—being open to new ideas—helps us remain competent and competitive as individuals, and in turn drives improvement and growth thinking within our teams and organizations.

> **Activity:**
>
> - Take time to reflect on these questions: Are you a learning leader? Do you have a growth mindset or a fixed mindset?
> - What is one step you might take to expand your curiosity and learning?
> - How might you engage your team in the process of being curious?

Utilizing Your Brain – Becoming Comfortable with Failure

Learning is also about becoming comfortable with failure, which is a potentially crushing eventuality of every great leader's journey. *"We need to learn and stretch and try things and by the way we'll probably experience some failures along the way. Naming that this is challenging is really important for creating psychological safety because once we name it, we kind of take off the table the otherwise implicit expectation that we're going to be perfect. We give ourselves permission to get it wrong"* (Amy Edmondson, speaking to Gagel, 2019 Greatness Podcast). Giving ourselves permission to fail—failing with grace and humility yet also resilience and resolve to persevere—is an important characteristic of great leaders.

> *"I stand before you as an 11-time world champion which makes me the most successful woman in the world at my profession. But it wasn't until I retired that I went back through all of my competitions and realised that I actually had lost a further 29 attempts at being world champion. I have lost more than I have won. Yet I'm the most successful in the world. For me to be successful wasn't about remaining undefeated. It was about how I handled moments of defeat because they are the more consistent feature of sporting life and indeed life. I had the privilege of sitting and listening to our first female prime minister Julia Gillard speak and she said that resilience is a lot like a muscle. You must keep using it for it to be strong and she's absolutely right in that. A muscle not used is a muscle that wastes and for us to be resilient, firstly, we have to accept the simple fact that we're not always going to win. We will actually spend more time losing, failing, being beaten, facing adversity and challenge, than we will experience moments of success."*
>
> (Anna Meares, speaking to Gagel, 2020 Greatness Podcast)

My philosophy about failure is that if I am not failing, then I am not trying hard enough and I am not pushing myself. Being a great leader involves taking risks, pushing ourselves to do things we may not believe we are capable of, and this can lead to failure. As author and speaker Sabina Nawaz shares: *"We all talk about how failure's good, it's going to teach us things. Let's learn. But who is going to say, 'Choose me to be your poster child for failure.' We're all trying to succeed, and life is messy, work is messy. Things happen and we cannot control it all. Unexpected things come along and there we are, we're failing. I would say if you're not failing enough, you're not experimenting enough. You're too much in your comfort zone and you're probably stymying future progressor are likely to get blindsided"* (Sabina Nawaz, speaking to Gagel, 2023 Greatness Podcast).

> As author and speaker Deborah Grayson Riegel shares: *"When it comes to failure one of the most important takeaways that we'd like people to know is that it is a normal part of life. It is expected. And we invite you to make a distinction between I failed at something, or the system failed me at something and as a result I didn't accomplish what I wanted to do, please make a distinction between that and I am a failure. The first one maybe has some guilt attached to it and the second one brings up feelings of shame, and while guilt can stop you or slow you down, shame can kill you. I want to make sure that people don't experience feelings of shame when something doesn't turn out the way that they want. It's not about you. It's not about your identity. You deserve to be in this world and on this planet and there is support for you."*
>
> (Deborah Grayson Riegel, speaking to Gagel, 2024 Greatness Podcast)

One step that can help is to prepare yourself for failure and create a culture where failure is an accepted outcome of learning and experimentation. *"How can you prepare yourself? Future proof yourself through experimentation, through learning, and yes, through failing. What if we could create some of that failure culture? And this is why I talk about starting small because we don't like to fail no matter what cliche we adopt. We don't like to fail. So how do you immunize yourself? How do you do it in little steps so that you can get that? You can build that capacity. You can build that muscle strength. But also, if you fail small, you're less likely to fail big"* (Sabina Nawaz, speaking to Gagel, 2023 Greatness Podcast).

You are perfect just the way you are. You will stumble. Hopefully, you will have a team surrounding you that helps you stand back up and brush off your knees. I encourage you to remove the fear of failure as an impediment to learning and growth. So go forth, learn, stumble, fail, learn more! Rejoice in new skills, new knowledge, and the fun of curiosity!

> **Activity:**
>
> - Take a moment to reflect upon an event that might have felt like a failure. What was your reaction? How did this event help you build resilience?
> - If you fear failure, who might you feel comfortable speaking with about this fear to help alleviate it?
> - How might you become more comfortable with the concept of failure?

Caring for Your Brain – Reflection

I head off to the beach each year (typically for a weekend in January) for my personal reflection and goal-setting time. I think about my purpose and values, my goals for the prior year and how I did, and my goals for the following year. I search for authenticity in my life. In 2021, I recorded how I spent every hour of every day for 365 days because I was not certain that I was voting for my values with my time, and then I headed off to Bermagui, a little fishing village on the coast of Australia, for nearly two weeks of reflection. No electronics—just me, the beach, and well, of course a few rounds of golf! And reflection.

> *"Sometimes you don't realize what you're learning when you're learning, and the most significant ideas and events that challenge and change and transform your thinking come often in random moments of quiet revelation when maybe you don't think you're going to learn anything at all."*
>
> (Amy Edmondson, speaking to Gagel, 2019 Greatness Podcast)

Carving out time to reflect—upon yourself as a leader, your team, and the world—is critical because reflection helps us understand ourselves as leaders and those we lead, and creates the downtime we need to fuel creativity and brilliance. *"Critical for anyone as they're exercising leadership, no matter what role you're in, is the need to pause, to get off the treadmill, to create some capacity, to build that muscle of reflection; whether that's to reflect on how you're showing up, whether you are holding yourself back due to some lack of confidence and need to build that back up again, whether you're not connecting in the right way and engaging your people in a really human way. No matter what it is, whether it's about confidence or connection, just the ability to create the pause in your life to breathe, to reflect, and then to set a new way forward, is critical"* (Michelle Sales, speaking to Gagel, 2020 Greatness Podcast).

Downtime is also vital for resilience—especially in today's technology-fueled abundance of over-connectedness. *"At the heart of it, with huge workloads and always-on technology and massive expectations and globalization and massive amounts of change, is that we are all under quite a lot of pressure. And that threat state is pretty much, for nearly all organizations and managers that I'm working with, an ever-present scenario. We have to be much better at self-leadership than we've ever had to be before and recognize that nobody's going to give us that time. Nobody's going to halve our inbox. We need to be able to manage that pressure effectively in order to not create bigger issues for ourselves and also behavioral issues that may then spread within the organization*" (Kristen Hansen, speaking to Gagel, 2019 Greatness Podcast).

> "*First of all, if you're going to block some time to reflect, block some time closely before that for 'to-do's' because if you go into reflection with a lot of 'to-dos', you're going to be very distracted. Secondly, do not call it 'just being time' or 'blank space.' Make it sound really important so if there are people who are looking at your calendar they don't think, 'Oh, what are they doing?' One of the reasons we don't do this is we're afraid of being judged. 'I feel so guilty sneaking away for this time because everybody else is working so hard', and you're working harder than them because you're doing the work that really needs to be done. If you gave yourself that blank space time, that brilliant idea will pop out. It's there in you, a gem waiting to be discovered and unearthed if you just give it the space.*"
>
> (Sabina Nawaz, speaking to Gagel, 2023 Greatness Podcast)

Making time for reflection—for downtime—is not being a "slacker" as a leader. It is about taking care of yourself like an elite athlete. It is about being a role model for your team. It is about showing up as the best version of yourself because you care enough to take care of yourself. Executive Coach and Co-Founder of the International coaching Federation David Goldsmith and I discussed this during our Greatness Podcast: Gretchen: "*When I was flying 120 flights a year I had a rule to never connect to Wi-Fi because my plane time was my thinking time, time spent journaling about 'how's the team doing' and 'what's the next conversation we need to have' and just literally thinking and reflecting about things. And then when I moved to Australia and stopped flying every week for a while, I had to be really intentional about figuring out how I created that reflection time without getting on a plane.*" David Goldsmith: "*What people don't realize is it looks hard to create the time, but that hour or two hours of thinking time has an exponential effect on the rest of the week ahead or the day ahead. It's hard to carve it out because it looks like it's not productive when there's so many things on the list to figure out, but we found that when you have that reflection time, and you deploy it regularly the leverage is amazing*" (David Goldsmith, speaking to Gagel, 2023 Greatness Podcast).

The feedback loops we create as leaders are also an important component of reflection. When I arrived in Australia in 2018, I created what I now call my

"Learning Journal," a different journal from the one I use for my day-to-day musings and reflections. I filled this Learning Journal with all the feedback I had ever received from profiles, HR evaluations, and many other different sources. Then, I created two pages that are important to me. One page, in blue ink, summarizes everything that is wonderful about me as a leader (that I am caring, tenacious, good at building relationships, etc.). Another page, in red ink, summarizes everything I can improve upon (that I can leave people behind, that I am at times not empathetic, and that I sometimes do not listen well enough to others' opinions). We all have superpowers, and we need to celebrate and read about these superpowers when we are having a bad day; additionally, we all have things we can improve upon as leaders. Gathering feedback from others is critical in this reflection process because we all have blind spots, and we all do things we may not realize make our teams and teammates crazy.

I believe in servant leadership (that to lead is to serve the needs of your followers and the organization), and this is a great starting point (Robert K. Greenleaf Center for Servant Leadership, n.d.). Decades ago, I started having my team review me anonymously twice a year. I begin the feedback process with each new team with two questions:

- What do each of you need from me as a leader to be successful?
- What does this team/organization need from me as a leader to be successful?

From the answers to these two questions, I create what a "good job" looks like for me as a leader, and in subsequent anonymous surveys, I probe into what I am doing well and where I need to improve. I believe anonymous feedback provides people with a safe space to say whatever they would like to say about you, and that if you have built a safe space with your team, they will say it. When I became CEO/President of WFCO, the team said things to me like, "We need you to help our organization learn how to better promote ourselves," and, "Each of us needs more clarity in what success looks like for our role." These were important ideas for me to hear, and informed my personal reflection as a leader.

As you gather feedback, it is important that you communicate with your team about what you heard, what you are going to work on, and what steps you are going to take to improve because of what you heard. I encourage you to pick only one or two things to work on at a time and to choose the skill that, if mastered, would have the greatest impact in making you the leader you aspire to be. For example, one year, I heard from my team that I had not given specific enough direction when I asked someone to do something for me (and this is an ongoing challenge!). I sat down with the team, agreed completely, suggested a few things that might help, they suggested a few things, and then I actively worked on it. Six months later, I asked, "Is this working? Am I getting better?" Teams really appreciate it when you acknowledge that you are not a perfect leader and are genuinely working to become a better leader.

Here is another great approach. My friend Christine Zeitz, retired Chief Executive and GM of Australia and New Zealand for Northrop Grumman, once shared a story that after a 360 review early in her career, she chose to work on not shutting down conversations too quickly during meetings with her team. Christine empowered a member of her team to observe her during meetings and privately provide feedback on how she was doing. Did she give voice to everyone? Did she let the conversation continue long enough, or did she shut it down too early? This is the sign of a great leader who can be vulnerable and ask for feedback.

Having the discipline as a leader to carve out time for reflection is essential. Building feedback loops into that process ensures that you are aware of key aspects of your leadership to reflect upon.

Activity:

- How much time are you currently carving out for self-reflection? Is it enough? If not, how could you carve out sufficient time to self-reflect?
- What are the key questions you should ask yourself during self-reflection? Here are some examples: "What are my priorities? Am I voting for them with my time? What have been my best moments as a leader? Some of my challenges? What do I need to focus upon improving?"
- How might you go about gathering feedback on your "blind spots" as a leader and effectively integrate this feedback into your reflection time?

Caring for Your Brain – Creativity

Back in 1999, when my son Holden and daughter Regan were two and three years old, I took an eight-week sabbatical—the first in the history of my organization. I asked for this sabbatical because I was burned out from traveling nearly every week, and I wanted my kids to know they were more important to me than my work. I was somewhat surprised that I returned from that sabbatical so refreshed and full of ideas, but research supports that downtime fuels creativity.

"I have probably literally asked more than 10,000 people in audiences when I do keynotes or small group workshops, I always ask where are the top five places where people are when they have their insights and by far the number one place that people yell out is the shower and the other four are driving or on any public

> *transport; in the middle of the night when your brain wakes you up with some great answer when you thought you were fast asleep; any form of exercise like walking or running or jogging. Then finally the last one that probably more men say that we do than women is on the toilet"* (Kristen Hansen, speaking to Gagel, 2019 Greatness Podcast). As Kristen goes on to explain, we have insights in these times because our brain is relaxed and is not "processing" data, and this frees up space for that great idea to fight its way through your brain clutter and declare itself.

Part of my challenge as a leader is that I am like a toddler: I run, and run, and run, and then I collapse. During my studies for my Master's in Nonprofit, I read another great book, *Sabbath*, about why we need REGULAR periods of rest to refuel our brains (Muller, 2000). Our prefrontal cortex is the computer system of our brain and controls our strategic and logical thinking (Hansen, 2018). I have learned from Kristen Hansen that the prefrontal cortex is an energy hog, and when we do not have regular downtime to refuel our brain, we are not operating at our peak performance as leaders (Hansen, 2018). Daily, we need to be taking breaks every 90 minutes; weekly, monthly, and annually, we need to invest time in fun activities and downtime outside of work.

Being creative as a leader and inspiring creativity in those you lead and influence can be challenging. It is not just about creating the downtime to fuel creativity; great leaders also create a safe environment for experimentation, risk-taking, and just "having a go."

> *"Take yourself to an escape room and challenge yourself to be in a dynamic environment where if you don't ask questions you're not even getting out of the starting blocks. There's a clock timer – 45 minutes or 60 minutes – and you've got to basically solve all manner of puzzles to work your way out of this room by the time the timer goes off. I'd sort of said quite arrogantly at this particular event, it was a party of eight-year-olds that was going on next to us, we were a whole bunch of 30-somethings, and I said, 'Oh gosh, do you have to make a lot of adaptations for the kids?' and they said' 'Are you kidding? The kids have every single game record in every single room we've got!' Firstly, they are much more prepared to ask for help. As adults we don't ask for help as readily as we should. The second thing is they're much more prepared to just have a go. They don't overcomplicate and overthink. They'll just start working their way through it to see if something works. They're a lot more experimental. So, I also say to people spend more time around young people."*
>
> (Holly Ransom, speaking to Gagel, 2022 Greatness Podcast)

Caring for your brain—creating downtime and space—are essential to your ability to remain creative as a leader. Without creativity, we lose our ability to generate unique solutions to the tough challenges we and our teams face. Creating that space for your team is just as important.

Activity:

- How effective are you at carving out regular periods of downtime to fuel creativity and innovation?
- How might you go about carving out 30 minutes a week as a starting point to take a walk without any pre-agenda, allowing your brain to roam free?!

Conclusion

Your brain is a powerful leadership tool. The ability to think critically, to learn, to fail, to learn again, to reflect, and to innovate is important and can help define you as a great leader. I encourage you to dive more deeply into these topics. Learn, fail, and enjoy the process!

"You were making me think of Sir Richard Branson, who I've had the great privilege of working with a bit through my career. He takes a notepad everywhere he goes. He's always taking notes. He is the most curious person in a conversation where practically everyone else in the conversation is just wanting to hear from him. He's desperately seeking new knowledge, and so I really think it is a superpower of people who sustain success because we're always having to be moving, learning, growing, particularly in the dynamism of the market as it stands right now. The half-life of skills is less than five years now, that notion that if you're standing still, you're falling behind has never been more real than as it is at this moment.

"Sadly, what we know is we all kind of follow this anthropological trend where we are really good at asking questions under the age of about five and six, peak question asks have sort of been defined to be four-year-old girls, that's like where you go if you want the most number of questions in a day, and sadly by the time we hit our teens, we're sort of down to less than double digits on a daily basis. We get this normed out of us for some reason, in part because I think so much about the educational model prizes perfection, getting the answer right. It stopped curiosity and experimentation and trial and error because that wasn't rewarded."

(Holly Ransom, speaking to Gagel, 2022 Greatness Podcast)

References

Baskin, J. (2022). 'The positive impact of creative outlets,' *MCI Institute Wellbeing Blog*, 31 January. Available at: https://www.mciinstitute.edu.au/wellbeing/creative-outlets [Accessed: 31 May 2024].

'Critical Thinking.' (2024). *Merriam-Webster Online Dictionary*. Available at: https://www.merriam-webster.com/dictionary/critical%20thinking [Accessed: 31 May 2024].

Dweck, C. S. (2007). *Mindset: The new psychology of success*. New York: Ballantine Books.

Gagel, G. and Edmondson, A. (2019). *Amy Edmondson discusses her book The Fearless Organization on psychological safety* [Podcast]. 30 October. Available at: https://open.spotify.com/episode/6joytIVhV3XlGFZx6rcodC [Accessed: 28 April 2024].

Gagel, G. and Goldsmith, D. (2023). *David Goldsmith, co-founder of the International Coaching Federation, discusses executive coaching* [Podcast]. 15 December. Available at: https://open.spotify.com/episode/1AktDJ16xXLuhPSr6ov1yG [Accessed: 2 May 2024].

Gagel, G. and Grayson Riegel, D. (2024). *Deborah Grayson Riegel discusses networking and women's relationship with failure* [Podcast]. Available at: https://open.spotify.com/episode/1eSA809JUvn8g9YfgSlkgt [Accessed: 15 July 2024].

Gagel, G. and Hansen, K. (2019). *Kristen Hansen discusses her book Traction: The Neuroscience of Leadership* [Podcast]. 29 July. Available at: https://open.spotify.com/episode/1ayV1cm3IedLOpHKA787Ka [Accessed: 28 April 2024].

Gagel, G. and Meares, A. (2020). *Anna Meares discusses leadership and team lessons from being an Olympic athlete* [Podcast]. 17 February. Available at: https://open.spotify.com/episode/3k1RqB8K3aiLVG5lRgkRol [Accessed: 28 April 2024].

Gagel, G. and Meyer, P. (2023). *Pamela Meyer discusses her book Staying in the Game: Leading and Learning with Agility for a Dynamic Future* [Podcast]. 1 December. Available at: https://open.spotify.com/episode/42CeGnq8Dx2DnbMadwhBd7 [Accessed: 2 May 2024].

Gagel, G. and Nawaz, S. (2023). *Sabina Nawaz shares insights on leadership* [Podcast]. 19 May. Available at: https://open.spotify.com/episode/7ac4m3RagAEiowVSU3Axrw [Accessed: 2 May 2024].

Gagel, G. and Nelson, B. (2019). *Dr. Brendan Nelson discusses leadership and personal values* [Podcast]. 10 May. Available at: https://open.spotify.com/episode/7cadZRiknfvUF63YeWcAlj [Accessed: 28 April 2024].

Gagel, G. and Ransom, H. (2022). *Holly Ransom discusses her book The Leading Edge* [Podcast]. 30 September. Available at: https://open.spotify.com/episode/0M5G6yRwqvbBlKa302wmKb [Accessed: 1 May 2024].

Gagel, G. and Sales, M. (2020). *Michelle Sales discusses her book The Power of Real Confidence* [Podcast]. 11 December. Available at: https://open.spotify.com/episode/79XJIyGSMH9TVysaGJjA8V [Accessed: 28 April 2024].

Hansen, K. (2018). *Traction: The neuroscience of leadership and performance.* Sydney: EnHansen Performance.

McGeorge, D. (2023). *The ChatGPT revolution: How to simplify your work and life admin with AI.* Hoboken, NJ: John Wiley & Sons.

Muller, W. (2000). *Sabbath: Finding rest, renewal, and delight in our busy lives.* New York: Bantam Books.

Ramsey, D. (2009). *UCSD experts calculate how much information Americans consume.* Available at: https://phys.org/news/2009-12-ucsd-experts-americans-consume.html [Accessed: 31 May 2024].

'What is Bloom's Taxonomy?' (n.d.). Available at: https://bloomstaxonomy.net/ [Accessed: 31 May 2024].

6

Telling Your Leadership Story

Introduction

When I work with a group of leaders, I often ask them within the first few hours to share what words come to mind when they think of me. They will often say things like "genuine," "passionate," and "knowledgeable"—someone once shared "eccentric!" They are never at a loss for words. First impressions really are a thing, and everyone has a brand; you just may not know what your brand is. My friend, author and speaker Gabrielle Dolan, has taught me a great deal about personal brand, and even more importantly, how storytelling is critical to enabling others to understand you as a grounded leader (Dolan, 2021). It's not enough to know what you stand for, to have courage, and to be able to use and rest your brain. How you take your authentic brand out into the world and tell your story are critically important to how you are viewed as a leader and how effective you are at building relationships. This starts with ensuring your voice is heard and that you are effective as a leader in lifting the voices of others.

> "It upsets me how so many people have so many dreams and so many ambitions and so much more potential but they're hiding, they're the world's best kept secret, they're afraid to speak up, they're afraid to share their dreams. My work is about building that belief because I know that unless we step into our individual brilliance, unless we own who it is that we are and who it is that we're being and who it is that we want to become, we are not ever going to be able to create that ripple effect of change."
>
> (Janine Garner, speaking to Gagel, 2022 Greatness Podcast)

Great leaders understand how to find their unique voice, as well as how to lift the unique voices of others. Meaghan Hooper-Berdik, Senior Vice President

New England with Turner Construction Company, shares this: "*Find your own voice. During the early part of my career I was watching how other people managed conflict, etc., and I imitated what I thought was powerful, strong, and effective. Then I found that I needed to be my own self, use my own voice, and much more effective for me was listening first, doing my homework, being prepared, being level-headed, a smart voice in the room. You don't have to be the loudest. Silence, listening can be impactful, powerful*" (Hooper-Berdik, 2023).

As you read this chapter, think about how you share your leadership brand—how you tell your story. What new ideas will you incorporate from the thoughts in this chapter? How will you ensure that people understand who you are as a leader? How will you tell your story as a grounded leader? How is the way you take your leadership brand out into the world important?

Having Voice

A friend says we always sit next to who we were meant to meet—and at a charity gala for the Denver Museum of Nature and Science, I did just that. I sat next to author, coach, and consultant Carrie Lynn Arnold, and as we realized we were both in the thick of our PhD studies, a friendship was born. Carrie later reached out to me to be a participant in her dissertation research study on executive women's voices being silenced at work, and I quickly agreed. As I prepared for my research interview, I felt certain I would have few stories to share about my voice being suppressed. I was shocked. In deeply reflecting upon my career in the construction industry during my research interview with Carrie, I realized there were numerous times when my voice had been silenced (either deliberately or unconsciously) by someone else, or even by myself.

One very concrete example occurred when one of my male colleagues took me to visit a client that we were transitioning over to me as Client Relationship Manager. My first meeting with the client was taking place the next day at a jobsite construction trailer, where we were meeting with the client team and one of their alliance contractors to talk about how the team could more effectively document the alliance's benefits. The night before that first meeting, I asked my colleague what he was planning to do to establish my credibility with the client and their contractor. I knew from experience that a woman's credibility is not typically established without a deliberate effort to do so. His response was, "Don't worry, I have this." A bit later in our preparation meeting, I asked him again for his plan to establish my credibility, and his response was the same.

The next day, we entered the jobsite trailer and shook hands with the entirely male group assembled, and then sat down at a long rectangular table—probably 10 of us. My male colleague launched into the agenda without introducing or acknowledging me in any way. I quietly sat through the two-hour meeting without contributing a great deal. As we were walking to our car, I turned to my colleague and asked, "So, your strategy for building my credibility with the client was to not introduce me and ignore me for two hours?" He was embarrassed and appalled by his behavior. It had not been malicious; he had just been so focused on the task at hand that building my credibility was not top of mind for him and he forgot. The good news is, I did establish my credibility with that client, and they became one of our largest clients.

I used this example with our team as a bit of a running joke to remind everyone of the challenges we face as women in the construction industry in having to continually prove ourselves in a way that men do not. It is important for our voices to be heard and respected, yet they often suppressed in both egregious and small ways. It was recently brought to my attention that being spoken over is one of the top microaggressions against women in the workplace, and I see it happen often.

"Our voices are suppressed through three different ways. The first would be a relationship is silencing us or we are feeling or experiencing the phenomenon of feeling silenced by another human. That could be a person in power, it could be a spouse, it could be a direct report, it could be a peer. It's not always a hierarchical structure nor is it always men silencing women. Women are silencing women in equal if not more egregious ways. The second way we lose our voice or feel muted and suppressed and muffled is through systems that silence. Sometimes we can take away all the people and replace them with brand new bodies, but there's a system in place that's designed to continue to honor certain things, suppress certain things. Sometimes when you can't quite figure out what's going on here, it could just be a system that is capable of silencing and that requires a little bit more analysis and investigation. The third way is we just do it to ourselves, we self-silence in the absence of someone else doing it. We often do it to our own voices and that can also create all sorts of health issues, and some would argue, well, everything is self-silencing. Everybody has a choice. We also must look at the socially constructed conditions in which we as humans are trying to operate and live within. My research was with women but I'm hearing many men come forward saying it's not different for me.

"Every single one of my participants in this research was a woman leader typically between the ages of 45 and 65, so these were women who were in power roles, had huge legacy of successfully running hospitals, running organizations, executive C-suite roles. To hear them discuss and talk about how even as a CEO they felt silenced, how they felt restricted in what they could or could not say, it's such a paradox. I think the theme I heard often was a feeling like they were choiceless and telling ourselves we're choiceless; and I think, also, the thing that

kept surfacing was questioning our own abilities. Do I belong? Do I deserve this seat? Maybe I'm not as good as dot, dot, dot, and so there's that rumination, and there's that saboteur voice. Being a woman in leadership is already so lonely. The percentage of women compared to men who are in those executive roles worldwide, we hold such a small percentage, and then to have one of those seats and feel marginalized in silence. That's not something we would necessarily want to share out loud.

"Sometimes we must deconstruct the entire phenomenon so that we can make sense of how it has impacted us. And sometimes we can't do that alone. Sometimes that is with a therapist, sometimes that's with our kitchen table colleagues. Sometimes that's with a coach. But sometimes we have to not just name it but then figure out how this has happened to us so that we can name it and declare it and then see it outside of ourselves, and then we can start to do additional work around it. This is not a light bulb moment. This is not necessarily flipping the switch and boom, my voice is there. The research suggests on average it takes about two years for a woman who has felt consistently silenced to get through that period of time. Sometimes it's faster, sometimes it's been a decade or more for certain participants to really get to the bottom of it and emerge from it. This is not a short journey, especially if we're mid-career, if we're in our fifties and sixties, this is deeper work, and you can't necessarily do it alone."

(Carrie Arnold, speaking to Gagel, 2022 Greatness Podcast)

30 years ago, I would not have even known what the phrase "to have voice" meant. Now I understand how we as leaders need to have voice, along with how we must give voice to others. When great people's voices are suppressed, they stop trying—or worse, they leave. Friend, author, and speaker Amanda Blesing shares these thoughts: *"One of the things that I observe that is backed by the research around ambition is that people say that women aren't as ambitious as men, and I say that's rubbish. What we know from the research is that men and women perhaps measure or value ambition differently. Men tend to correlate success and ambition with financial gain. For women, when it comes to being ambitious, the money is important but if women don't feel like they have a voice and they're being heard and if they don't feel like they're actually making a difference, women are far more likely to pick up their bat and ball and leave or take their foot off the accelerator and not try as hard anymore"* (Amanda Blesing, speaking to Gagel, 2019 Greatness Podcast).

Giving voice to ourselves as leaders requires being present to our voice—being present to whether our voice is being silenced in some way by a person, a system, or ourselves. The next step is to examine why, perhaps speak with someone about whatever is happening, and stand up for our right to have voice. Empowering others to have a voice requires being mindfully present in conversations and meetings, and deliberately considering how each person is experiencing the interaction. One might even ask, "Do you feel your voice is heard?" or "Did you feel you were able to share your thoughts in this conversation/meeting?"

To take a person's voice is to rob them of their dignity. Respect and voice coexist and are critical to our relationships at work and at home. Our ability to recognize when we are losing voice and to do something about it (as well as to recognize when others' voices are being suppressed and do something about it) is a sign of great leadership.

Activity:

- Think of a time when your voice has been suppressed. How did that make you feel? What might you have done differently?
- What can you do to be more present to ensuring that everyone's voices are lifted?
- During upcoming meetings, reflect upon how people either have or do not have voice. What can you do to help lift people's voices?

Your Leadership Brand

Early in my construction industry management consulting career, I vividly remember being on a plane with my boss, Lou Bainbridge, who turned to me and asked, "How are you going to show up to your clients?" I was stumped. What did he mean? Lou asked me to think about it, and I did. This was my first conscious attempt at considering my brand, and how others would experience me. I began thinking about how I wanted to show up—words like "helpful," "professional," and "confident."

In her (2023) book on personal branding, *From Invisible to Invincible,* Amanda Blesing defines your personal brand as a strategic approach to ensuring your unique skills, your results, and your value are known in professional settings. Her concept emphasizes authenticity and intentionality in showcasing your strengths, value, and career aspirations. She stresses the importance of visibility, confidence, and networking, encouraging women to break through societal barriers and actively promote themselves to achieve their career goals.

I encourage you to try this exercise. Ask five people what words pop into their mind when they think of you as a leader. Or better yet, ask someone else to ask five people this question. Now ask yourself, what five words do you WANT people to use when they think of you as a leader? Are these the same words? Different? This is not about disingenuous packaging to promote yourself as a great leader; this is about how you want to show up in the world in a way that supports your purpose and values, and behaving in a way that helps others experience you that way. Your personal brand—your leadership brand—is how you show up as a leader. How do people describe you as a leader? Do you create psychological safety? Are you adept at setting direction for the team? Do you boss people around? Do you shrink from decisions? Do you stand up for what you believe in? Who are you as a leader, and what do you value?

Thoughts on personal brand and the image we portray at work arose often during my interviews for this book when I asked what advice my interviewees would give women entering our industry:

Anne Ramsey, Director Global Project, Construction, Facilities, and Utility Management, Procter & Gamble: "*Be humble and ask questions; own your craft – safety, mechanical, quality; listen to understand; be visible and be aware of the image you portray, the perception you give off.*"

(Ramsey, 2023)

Brian Gallagher, Vice President Corporate Development, Graycor: "*Be clear about your goals, your path, and how you'll develop yourself; take advantage of educational projects, say yes and challenge yourself; be willing to do the hard stuff; build your personal brand and reputation by doing quality work and following through; look for people who can advocate for you and build rapport with people.*"

(Gallagher, 2023)

Josephine Sukkar, Founder Buildcorp and Independent Director: "*Value experience and don't run after the job title, run after the experience; commercialize your brand and build your CV.*"

(Sukkar, 2023)

I recently attended a dinner with 20 women participants of the APGA Women's Leadership Development Program and seven members of that organization's board and staff. When I looked at the three most successful women in the room, I realized that we all have "big personalities." We are confident, perhaps louder and more outspoken than most women. It was another light bulb moment for me. Most of us who have thrived in the past 30 years in the construction industry HAD to be big personalities. I am grateful that as an industry, we are beginning to understand that it is not just the loudest voices in the room—the biggest personalities—who have what it takes to thrive in our industry. We are becoming more accepting of the fact that leadership takes many forms and that there are many different types of leadership brands that can succeed in the construction industry.

Understanding your unique leadership brand and owning it is critical to living a life that is authentic to you. There is no one right answer to what your leadership brand should be. It must be authentic to you and reflective of your personality, your values, and your style.

Activity:

- Write down the five-to-ten words you would like people to use to describe you.
- Ask a few people what words they WOULD use to describe you (or ask someone else to do that for you).
- Reflect upon the outcome of this exercise.
- Is there a word you would like people to use to describe you that they are not currently using?
- What one or two things could you do to more strongly convey that value—that element of your leadership brand?

Storytelling

I was sitting at an energy conference listening to the female CEO of a gas utility speak about safety. She could have started with slides featuring safety statistics or information on their safety program. Instead, she told the story of having to knock on the door of a family's home to tell them their husband and father had died because he did not follow protocol on an emergency response, and looked into a hole in the ground before his safety partner had arrived at the scene. That story, and the details she shared of that day, brought home the need for safety protocol in a way that riveted the audience. That story conveyed her values and those of her company, as well as the message that we must do better. That story created an emotional connection with the audience that I will never forget.

"I think sharing personal stories is one of the critical ways you can really build relationships with your team and your customers to demonstrate authentic leadership. Most leaders start by sharing not personal stories but perhaps stories about a work situation and it can end up being more of a case study than a personal story; or they might share stories about people they admire, so they'll share stories about Nelson Mandala or Steve Jobs. It's when you have the courage to share a personal story about something that happened to you on the weekend or about a lesson you learned when you were growing up or getting told off by your mother when you were eight, these personal stories reveal something about you. That's when it can not only fast-track relationships and trust and respect, but it can also deepen relationships which again makes it a really, really powerful communication skill and leadership skill. By sharing that story their people say, 'Oh, they're

real, they're human, they have the same problems we have.' I love the work from Brené Brown that she's done on vulnerability. It's the one thing we're reluctant to show, vulnerability, but it's the one thing we really respect in other people when they do it."

(Gabrielle Dolan, speaking to Gagel, 2021 Greatness Podcast)

A few years ago, I was asked by the American Gas Association to speak to their women's leadership group about personal branding, and I utilized Aristotle's Modes of Persuasion—ethos, logos, and pathos—as the framework for the discussion ("Understand the difference between ethos, pathos, and logos to make your point," 2022). Ethos is the credibility and trustworthiness we demonstrate by declaring our educational background, titles, and experience. For example, I might tell you that I have successfully completed a PhD in leadership and worked in the construction industry for 40 years to establish my brand and credibility. Logos is the logical reasoning we utilize via value statements that articulate the things we have achieved. For example, I might say that I have successfully run a women's leadership program where 90% of the participants have reported an increase in confidence. Although these first two strategies are important foundational elements of our story, it is the power of pathos (how we connect with people's emotions) that brings about strong relationships and genuine understanding of one another. We often do this by telling stories that create an emotional connection.

To demonstrate ethos, logos, and pathos, I might introduce myself in three ways. In the first introduction, I list off my degrees, titles, and qualifications; in the second, I might focus on my measurable achievements with clients and leaders in the construction industry, like helping an energy company reduce their capital construction underspend from an average of 14% per year to less than 1% per year; and in the third, I share my earlier story about how at the age of 17, my father told me, "Girls don't do engineering," and how this experience fueled my passion for women's achievement in our industry. All three approaches have value. Which introduction is most compelling to you?

Many of us are technical people who prefer to stick to the facts and avoid emotions. But emotions help create connection, convey our values, and articulate our leadership brand. It is not enough to understand your leadership brand. Your ability to humbly share your superpowers as a leader with others—to speak your leadership brand to the world—is just as important. "*You do amazing work, and your results should speak for themselves. But unfortunately, in a really busy world, our results don't speak for themselves, and we need to learn to speak to our own results, to tell people about the outcomes that we delivered; and we need to learn to do that in a way that's congruent and people don't think we're bragging.*" (Amanda Blesing, speaking to Gagel, 2019 Greatness Podcast).

Stories are not bragging. Stories must be genuine and convey the essence of who you are. Individual storytelling is about being able to genuinely connect with people in a way that causes them to understand what we stand for, what we value, and what we hope to achieve. The stories I am talking about are not boastful stories or bravado; they are the genuine stories of our experiences that shape our lives and convey our values and accomplishments to others.

I once asked a group of Australian women leaders to create "value statements"—how they articulate the value they add to their team and organization. After our meeting, I was walking to dinner with one woman who had struggled a bit with the exercise. I asked her about her recent accomplishments, and she proceeded to tell me about an award her construction project team had recently won. When I asked why she had not used this example in the value statement exercise, she said, "Well, it really wasn't me, it was the team..." Baloney! We can claim our part in team success! I suggested she start with something like, "I'm proud of the contribution I made to our team's success and happy about the recognition we received via this award. I played a role in helping us achieve strong community engagement by..." This is not bragging or arrogance. It is claiming your worth through a story.

> "*I'm a big believer in a few strategies including the humble brag formula which is a little bit like an elevator pitch. To do the humble brag formula you need to actually be collecting regular achievements, catalogue your achievements, and be able to talk about your achievements in a way that takes people along for the journey. Maybe it is a storytelling model or maybe it just brings people along on the journey a little bit so that when you actually land the result, it's not just saying, 'Hey, look at me, I won six awards.' It's like, 'So, last year I was working on this particular project. It was really interesting, fascinating, had these three elements. And part of it was bringing the team along and I had to do this and am delighted to announce that as a result, the company was able to deliver this result.' That doesn't sound like bragging, does it? That just sounds like, 'Wow, I'm really delighted to announce that ...'*"
>
> (Amanda Blesing, speaking to Gagel, 2019 Greatness Podcast)

The ability to effectively tell stories to convey your value, motivate your team, and inspire others is a skill that must be cultivated just like any other leadership attribute. "*I think one of the things you probably should start with is just trying to identify all the stories that you could potentially use, and I think it's a good idea to do this without having a message in mind because it will free you up a bit. I suggest you get a blank piece of paper and give yourself a good hour or so to do this. From your earliest memory write down everything you can remember. Normally people will go through that quickly and just pull out the big events in their life. But if you sit with it for a while, you'll be surprised how many random things come into your mind. Then the following week you might go, 'I've got to do a presentation on innovation, I need a story on innovation, what it means to me.' And you look at your list and you say, 'Oh, actually I could use that, I could use this story or that story'"* (Gabrielle Dolan, speaking to Gagel, 2021 Greatness Podcast).

> I was recently working with a group on storytelling, and one woman in the group stated that she really did not do anything of value. I asked what her job was, and she said, "Project Sourcing." "What does that mean?" I asked. "Well, it means that I make sure all of the people and materials are in place to keep the project running." That sounds like something of value—a story worth telling!

The first time I gave a talk about my book on working moms, I included a picture of me as a baby sitting on my mom's lap. I told the story of my stay-at-home mom and how no matter what crazy decision I made—changing jobs, a divorce, traveling with young children—she was always my biggest fan. Then I began to cry. It took me a moment to regain my composure, but you could see on the faces of the women in the audience that this story had touched them in a meaningful way, and that they understood my message: that we all need to have raving fans who support our decisions. So, I kept that slide in for future talks.

> *"It was like sharing a story and putting a picture up there of your mum. Of course, you're going to get emotional every time you tell that story, as you should. And that is what that does it, it creates a real connection with your audience and the people you're speaking to."*
>
> (Gabrielle Dolan, speaking to Gagel, 2021 Greatness Podcast)

I realize it may not be easy to show emotions in the construction industry. We have constant reminders to keep our emotions at bay. It is okay to share your authentic brand, to tell the stories that help convey vulnerability and who you are as a grounded leader.

Telling our stories as leaders—sharing our leadership brand—creates real connection and authentic, genuine leadership. Waiting for others to tell your story, to share your accomplishments, is sometimes a long wait. It is okay to have pride in what you do, share how you contribute to the success of your team and your organization, and tell those stories in a way that connects with the emotions of others.

Activity:

- Take a moment to think about your values and your superpowers.
- What stories could you tell me that convey the essence of these values and superpowers?
- How might you share the results of your hard work in a way that is not bragging, but definitely conveys the value you contribute to your team?

Conclusion

Your ability to understand and thoughtfully share your leadership brand with others in a genuine way that creates connection—your ability to have voice and to lift the voices of others—will set you apart as a leader. I encourage you to claim the space you want to occupy in our industry, be proud of your superpowers as a leader, and reflect upon how you will leverage storytelling to convey your leadership brand.

"Say I go into a senior leadership team that have been working with each other for years to work on the process of helping them share personal stories to communicate their company values or the new strategy or the new purpose and that's the reason for the workshop and the reason for the training. At the end, everyone says,' I feel so much closer to everyone in the room, and there'll be this constant message, 'I've worked with you for five years and I did not know that about you'. The connection is so much stronger within that team because they've shared stories."

(Gabrielle Dolan, speaking to Gagel, 2021 Greatness Podcast)

References

Blesing, A. (2023). *Invisible to invincible: A self-promotion handbook for executive women*. Victoria, Australia: BookPOD.

Dolan, G. (2021). *Magnetic stories: Connect with customers and engage employees with brand storytelling*. Hoboken, NJ: John Wiley & Sons.

Gagel, G. and Arnold, C. (2022). *Carrie Arnold discusses her book Silenced and Sidelined—How women leaders find their voices and break barriers* [Podcast]. 19 August. Available at: https://open.spotify.com/episode/0QnMQ73dUBxPBJNk OQEM3V [Accessed: 1 May 2024].

Gagel, G. and Blesing, A. (2019). *Amanda Blesing discusses her book Invisible to Invincible on personal branding* [Podcast]. 16 September. Available at: https://open .spotify.com/episode/2tCXZYQXAfTVMIRufeNLTb [Accessed: 28 April 2024].

Gagel, G. and Dolan, G. (2021). *Gabrielle Dolan discusses her book Magnetic Stories* [Podcast]. 2 February. Available at: https://open.spotify.com/episode/ 0uamofsvN18F1sWcjawMQm [Accessed: 28 April 2024].

Gagel, G. and Garner, J. (2022). *Janine Garner discusses her book Be Brilliant* [Podcast]. 14 October. Available at: https://open.spotify.com/episode/ 2jTwgPFgU2jmhi9kG6WNTW [Accessed: 1 May 2024].

Gallagher, B. (2023). Interview by Gretchen Gagel [Zoom]. 25 October.

Hooper-Berdik, M. E. (2023). Interview by Gretchen Gagel [Zoom]. 20 November.

Ramsey, A. (2023). Interview by Gretchen Gagel [Zoom]. 16 November.

Sukkar, J. (2023). Interview by Gretchen Gagel [Zoom]. 21 December.

'Understand the difference between ethos, pathos, and logos to make your point.' (2022). Available at: https://www.thesaurus.com/e/writing/ethos-pathos-logos/ [Accessed: 31 May 2024].

Part II

"Me" – Grounded Self-Leadership – Conclusion

I believe the foundation of Grounded Self-Leadership begins with the purpose and values that are unique to you. We have only ourselves to be, and each of us is a unique leader on a journey of self-discovery. Our ability to have courage and build our own confidence is a critical component of our effectiveness as leaders. How we utilize and care for our brains (the most amazing tool we have to think, be creative, and solve problems) helps define who we are as great leaders. The ways in which we live our authentic leadership brands and tell our stories to create emotional connections helps others understand who we are as great leaders. Be true to who you are as a leader, and greatness will come.

> "When I was offered a position to be the flag-bearing team captain at my final Olympics for Australia, I thought I was going to have to change how I led. How do you lead the most motivated group of people you've ever come across in your life? And my father simply said to me, 'You don't have to change a thing because you've led yourself and your life in the way that they want you to lead a team.' I didn't realize that. I thought leadership was a very different concept. But I had been leading my whole life through my actions."
>
> (Anna Meares, speaking to Gagel, 2020 Greatness Podcast)

Reference

Gagel, G. and Meares, A. (2020). *Anna Meares discusses leadership and team lessons from being an Olympic athlete* [Podcast]. 17 February. Available at: https://open .spotify.com/episode/3k1RqB8K3aiLVG5lRgkRol [Accessed: 28 April 2024].

Part III

"We" – The Supporting Cast – Introduction

I was sitting at an IWFA dinner in Melbourne one night speaking with a friend about my coach, Edgar Schein. I remember her asking me, "How did you know you needed help?" I probably responded that I was feeling overwhelmed with everything on my plate, and given that Edgar had many similar roles in his life (and he is brilliant), I thought he might be able to help. It seemed natural to reach out to Edgar. Our conversation broadened out to include all 10 or so of the women sitting at the table as my friend asked the group how they knew when to ask for help. I distinctly remember one woman's response, that she did not think she had ever asked for help in her lifetime of 60+ years. I thought to myself, "Either she thinks she has all the answers, or she's too shy to ask." Either way, I was surprised.

Early in my career, I spent little time deliberately thinking about having a mentor, sponsor, coach, or ally. "Networking" was not a term in my vocabulary. When I showed up at my first Ralston Purina plant in Oklahoma City, I was in survival mode. Show up, do a good job, and try not to fold under the pressure. Thankfully, nearly 40 years later, I am grateful for the many mentors, sponsors, coaches, allies, and networks I have experienced and continue to develop. I call these people "Team Gretchen," or my "raving fans," and that team has changed over time. My good friend Janine Garner wrote a book about it: *It's Who You Know: How a Network of 12 Key People Can Fast-track Your Success* (Garner, 2017). These are the people who cheer me on when I am winning, and pick me up and dust me off when I take a tumble.

Sometimes, it is hard to ask for help. We might think everyone else is doing it on their own, and that asking for help is a sign of weakness. I believe asking for help is a sign of strength, however—a condition of being a great leader. To ask for help is to use intelligence. I had never completed a PhD before. Wouldn't it make sense to turn to people who had to ask for advice and mentorship? You are not the "teacher's pet" if you have a sponsor within your company. You are a savvy businessperson who is using a resource to advance your career, as many people do. Thinking strategically about your "supporting cast" is smart and fun. I have loved every minute I have spent with these people in my life.

As you read through this portion of the book, I encourage you to reflect upon your current support network and the steps you can take to strengthen that network while also building enduring friendships along the way. Rarely are we successful alone, and cultivating your "supporting cast" is important as a leader.

Reference

Garner, J. (2017). *It's who you know: How a network of 12 key people can fast-track your success.* Hoboken, NJ: John Wiley & Sons.

7

Learning from Mentors, Sponsors, and Coaches

Introduction

When I first heard Australian Olympian Anna Meares speak at an APGA conference, I distinctly remember one slide in her presentation. It began with a picture of Anna in the center of the slide. Then, one by one, a picture of each person on her team appeared around her (her nutritionist, her psychologist, her trainer, etc.). Anna described the vital role each person played in her success and drove home the point that we see her on the Olympic podium as an individual competitor, but none of it would be possible without the members of her team and the incredible support each person provides.

> "*I can't do it on my own. Even though I'm an individual athlete, it takes a great team of people who are all experts in their field. They dedicate their life to being an expert in areas such as nutrition, strength and conditioning, mechanics, biomechanics, science, sports science, statisticians, coaches, tacticians, skill acquisition specialists, media, analysis, management, administration staff. I can't be an expert in each of these fields and who fills each role is critical to not just the team's success but the outcome for me as an athlete. Each person is like a spoke, and you need spokes that pull from opposite sides of the hub to make that wheel function at its best capacity. You need to be challenged. You need people with different ideas, with different thought processes. Even if it's just to reiterate to you that you're on the right path, at least you're thinking from every perspective possible. I never blamed anyone for a loss, and I wore a loss very heavily, but I thanked every person I possibly could whenever I got a win because I knew that it wasn't just on me.*"
>
> (Anna Meares, speaking to Gagel, 2020 Greatness Podcast)

Building Women Leaders: A Blueprint for Women Thriving in Construction, First Edition. Gretchen Gagel.
© 2025 John Wiley & Sons, Inc. Published 2025 by John Wiley & Sons, Inc.

Mentors, sponsors, and coaches are three important categories of people who support your success as a leader, and each plays a vital role in helping you thrive, develop personally, and achieve your career aspirations. These are people you trust, you admire, and you respect. These people believe in you. *"I remember an exercise that someone asked me to do a long, long time ago which is to think of someone who made a difference in your life. It could be an early boss or a teacher. Most people report thinking of someone who saw something in me I didn't see in myself. A person who cared enough to see something worth developing, worth unleashing. They believe I'm able to do more than I know and as a result I'm inspired to live up to it"* (Amy Edmondson, speaking to Gagel, 2019 Greatness Podcast).

As you read through this chapter, think about the people you admire and who believe in you. Think through the different roles they play (or could play) as mentors, sponsors, or coaches, and how you might strengthen this supporting cast.

Mentors

After my first two years working in a pet food plant, I was frankly fed up. I was tired of the wind in Oklahoma (yes, the wind does come "sweeping down the plain!"), tired of my neighbors' dogs licking the beef tallow off my steel-toed boots in the communal mail room, and tired of feeling alone and constantly challenged. I told the Plant Manager, Mark Burns, that I had taken the Law School Admission Test (LSAT) and that I was, in a phrase, "out of here." Mark was Ralston Purina's youngest Plant Manager and empathetic to my frustrations. Instead of lecturing me, Mark told me how much potential he saw in me and how valuable I was as a trailblazer for women at Ralston Purina. He asked me to stay with the company and transferred me to Davenport, Iowa, to run a cereal plant. It was a chance to step up and run my own show, and my first experience of being mentored (and probably sponsored) by someone.

A mentor is "a trusted counselor or guide" ('Mentor,' 2024). As a trusted guide, a mentor shares invaluable advice with you in both your personal and your professional lives. Trust is important, as mentors are only valuable if you are able to share your fears, your challenges, and your concerns. Another of my early mentors, Dub Bankes, was someone I could turn to when I did not have the answer or I faced a challenge that seemed insurmountable. Dub had a wonderful sense of humor that set me at ease, and he never made me feel stupid for asking him questions.

> *"Looking back, I wouldn't be the person I am without mentors and coaches, people who pulled me forward when I didn't want to go forward. I encourage men in our industry to look for opportunities to give women experience, exposure, and leverage their strengths."*
>
> (Ramsey, 2023)

I currently have four wonderful mentors in my life—people I turn to for advice on any number of topics—and I am able to speak to each of them about different things. Some provide guidance on my personal life, and some more so on my professional life. Sometimes, I am not looking for advice, and just need someone to talk to in order to work through my decision. Great mentors, like great leaders, ask great questions that help you think about things from different perspectives and provide new insights. Mentors rarely give you the answers. They ask you the questions that give you the answers.

> As Maddie Gosser, Human Resources OpX Program Manager for Baker Construction, shares: *"Mentors have been a huge part of my entire career, a sounding board. They are there when I don't have the lived experience and they help me navigate the challenges. They are cheer leaders. If I mess up, I can ask, 'how do I fix this?'. They are my own board of directors."*
>
> (Gosser, 2023)

The people I interviewed discussed the importance of men mentoring women in our industry:

Kim Neuscheler, Vice President and General Manager, Turner Construction Company: *"Most of my mentors and advocates have been male as they are the majority of the leaders in our industry. I have leaned on mentors to teach me the ropes, to call me out on stuff. Tom was always fighting for me, saying, 'Kim can do it', always willing to challenge me and give me real time feedback, to help me grow as a leader. We all need someone like that in our corner."*

(Neuscheler, 2023).

Andy Browning, General Manager of Engineering and Construction, Duke Energy: *"I encourage men in our industry to be a mentor, to help* *show women how to do things that they are new to, to push them and not let them stay in their comfort zone, to stretch them. I put women somewhere that they will succeed and get their first win, someplace they can hold their own and build confidence."*

(Browning, 2023).

Kelcey Henderson, President and Co-Founder, Continuum Advisory Group: *"My advice to men in our industry is to serve as a mentor, an advocate, a peer. Use your male voice to advocate for women, lift up their ideas, give credit where it's due. When you see women doing work behind the scenes, celebrate that work, recommend them for that promotion."*

(Henderson, 2023).

Jayne Whitney, Chief Strategy Officer, John Holland Group: *"I wasn't brave early in my career. Good leaders gave me the space to grow and develop. I encourage women to be discerning, work for good people who believe in you more than you do and seize the moment."*

(Whitney, 2023).

Tracy Gu, Vice President of Cockram Construction USA: *"Rob is a strong mentor. He trusts me and gives me the flexibility to use my intelligence to manage the business."*

(Gu, 2023).

When looking for mentors, I encourage you to think about the advice that might be most helpful to you at this point in your life. Is it advice on how to do your job? How to advance your career? How to juggle work and personal responsibilities? How to dig yourself out of burnout? Then, consider the people you know within and outside of your company who have knowledge and advice related to these important questions. You might consider these questions as well: "Who are the people I most admire? Who are the people that most admire me? Who are the people with the biggest hearts and compassion? Who are the people I trust?" Mentors might be a level above you or five levels above you, in another company, or not working at all.

The next most frequent question is, "How do I approach a mentor?" You could take them out for coffee or lunch prepared to have a conversation like this:

> "Susan, I really appreciate you taking the time to meet with me. I have so much admiration for you because of your career accomplishments, your problem-solving skills, and your ability to build relationships with people in our company. As I continue to progress in my career, I'm cognizant of the fact that I have much to learn, and I know that mentorship is an important way to

> continue my development. I'm looking for a mentor who would be willing to meet with me quarterly for an hour to provide advice and guidance. Would you consider being my mentor?"

This is one approach, and yours will be unique to you and the person you are considering as a mentor. You might also have informal mentors in your life, people you have not had this direct conversation with but who you know you can go to for advice as needed. They may not even realize you consider them to be your mentor, and that is okay.

Most importantly, do not be afraid to ask someone you think is well beyond your reach. As Chair of the Board of a construction company, I informally mentored a young project engineer after meeting her on a construction project visit. Afterward, she approached me via email to see if I could provide some guidance on how to become more involved in the construction industry, and we now meet regularly. It is as helpful to me as it is to her because she brings a fresh perspective to my thinking.

There are also external mentoring programs to consider when looking for a mentor. I participate in a globally expanding program in Australia called "Mentor Walks," founded by my friend Bobbi Mahlab. Each first Friday of the month, a group of women mentors and mentees gathers for a one-hour walk around the beautiful Melbourne Royal Botanical Gardens. The program is effective because of its simplicity and the careful curation of matching each mentor with two-to-three mentees based on their unique questions. I enjoy not only the mentoring role I play but also learning from each of the wonderful women I meet.

> *"It's a simple model that's having a dramatic impact. We know that 94% of the women who attend our walk say that they received the advice that they came for and that speaks to the curation of mentees and mentors. We know that 33% of the women who attend our walks have started new roles and say that Mentor Walks was an important factor in their decision; and that almost half the women who have come said that the conversations they've had helped them move forward. Many industry and professional associations also have mentoring programs. I would investigate several mentoring programs before deciding upon one that seems to meet your needs."*
>
> (Bobbi Mahlab, speaking to Gagel, 2023 Greatness Podcast)

Bobbi's last point is important. Many industry and professional associations offer mentoring programs, and taking the time to investigate several options before selecting a mentoring program is an important upfront investment of time. I also encourage you to be a mentor, as this is a great way to pay it forward.

Here are some additional tips regarding mentoring:

1) If your mentoring relationship is not working for you, do not feel bad about discontinuing it. Politely thank your mentor for the time they have spent with you and indicate that you think it is time to move forward with advice from other mentors.

2) Sometimes, your mentor might not have time for you, and that is okay too. If this continues, again, politely thank them for the time they have spent with you and indicate that you think it is time to move forward with advice from other mentors.

3) There is no "correct" number of mentors. It is what feels right for you. These can be both formal mentoring relationships and informal relationships.

I cannot imagine my life without mentors—especially as I have transitioned to different roles and industries. These people are a source of guidance and support that enriches their lives and mine alike. What a gift! I encourage you to thoughtfully consider how you can start or continue involving mentors in your leadership journey.

Activity:

- Think about where you are in your career and in life. What type of advice would be helpful to you?
- Brainstorm a list of people you admire who might be able to provide advice in these areas.
- Think about the criteria you might use to evaluate who is a good fit for your needs and rate each person on these criteria. I use a simple "1, 2, 3" approach—"low, medium, high"—for ranking.
- Is there at least one (or possibly two) people you might approach to be your mentor? Or informally ask out for coffee or lunch for advice?
- Write down how you will approach them, and remember to be specific on the "why" and what you hope to gain from the relationship.

Sponsors

One day during my time leading the cereal plant in Iowa, I was challenged by my boss's boss, Ron Erps, in our weekly staff meeting about the cleanliness of my plant. I was devastated. Then, I took ownership and my team whipped our plant into an even higher level of cleanliness. What I determined later was that Ron was testing me, for both my leadership abilities and my resolve. Ron wanted to see whether I would rise to the challenge or crumble.

About two years later, Ralston Purina purchased Beech-Nut Baby Food from Nestle. Beech-Nut had been convicted two years earlier of 215 felony counts for shipping baby apple juice that was labeled as "pure," but in fact was a mix of cheaper juices and sugar syrup (Buder, 1987). At the time we purchased the company, it was hemorrhaging money. Ron was selected as the head of operations for Beech-Nut, and about a week after his selection, I walked into Ron's office and asked to go with him. I felt this might be a tremendous learning opportunity and that Ron would continue to be a mentor I could learn from. Ron said "yes."

I did not have an official position for about the first six months. Instead, Ron asked me to "find problems and fix them." It was a tremendous learning experience for me as I uncovered all types of issues, from gender-based daily employee lay-offs to a plethora of unsafe practices, to large amounts of ingredient waste. Then, Ron put me in charge of manufacturing in Canajoharie, New York, where, as a 26-year-old, I was managing a $28 million operating budget and hundreds of employees. We began turning a profit in nine months, three months earlier than the deadline Jim Nichols, our President, had set for us.

I did not understand this at the time, but not only was Ron my mentor, he was also my sponsor. A sponsor is defined as "taking responsibility for some other person or thing" ('Sponsor,' 2024). I can only imagine the conversations (perhaps arguments) that preceded my appointment to that operations position at such a young age. Ron took a chance on me, advocated for me, and for that I am grateful. A sponsor is typically someone within your organization because they are advocating for your advancement professionally, and that is difficult to do from outside of the organization. A sponsor understands your aspirations and provides you with advice on what skills you might need to build or on how to navigate the political landscape of your organization to achieve these goals. Sometimes, sponsors will challenge you, asking you to take on responsibility you might not feel you are ready for. They say great things about you when you are not in the room and provide you with visibility within the organization. They believe in you and want to help you succeed.

> *"Research by Sylvia Ann Hewlett at the Center for Talent Innovation shows that cultivating your personal brand is one of the best ways to attract a sponsor – and professionals with sponsors are 23% more likely than their peers to be promoted."*
>
> (Clark, 2018)

Those I interviewed for this book reinforce the importance of sponsors:

> *"The late David Clark, founder of Macquarie Bank, put me on my first board, and was so lovely and set me up to succeed. He even rang me the night before my first meeting to see if I had any questions. Now I'm Chairing the Australian Sports Commission and David would be proud."*
>
> (Sukkar, 2023).

> *"I encourage men in our industry to sign up to be a coach, a sponsor, to be willing to work with women, take risks, put them in positions to learn, grow, and demonstrate capabilities."*
>
> (Ellis, 2023).

As you search for the right sponsor, I encourage you to look up one, two, or even three levels in your organization. Think about who knows you and your skills, or who might be open to learning more about you, your skills, and your career aspirations. Who has political influence within the organization? Whose opinion of you will be heard and respected? Who is approachable and will help you think boldly? Who has sponsored other women and men within the organization?

You might approach a sponsor by reaching out to them in this way: "I've been thinking about my career aspirations here at XYZ Company and I'd love to pick your brain about this topic. Would it be possible to meet for 30 minutes over coffee?" Then, the conversation might go something like this:

> "Heidi, as I mentioned, I've been thinking a great deal about my career path here at XYZ Company. While I'm really happy doing what I'm doing now, I could envision myself in five years either as a General Manager or as a Vice President in Department X or Department Y. These positions interest me because I feel like I could really leverage both my current skillset and the skills I'll have gained over the next five years to add value to the organization. I've read quite a bit about the value of having a sponsor within an organization, someone who can help me understand the skills I need to gain to achieve these positions and help me be considered for different roles that broaden my skillset. Would you have an interest in being my sponsor?"

This is just an example, and your request could be accomplished in many different ways that fit your style. If someone says "no," do not feel bad; try again with someone else.

As I was writing this book, I unfortunately heard a story which underscores that it is not enough to sponsor someone into a position. You must also continue fostering their success. *"He put me into a new position with more responsibility as my sponsor. A year later HR called and said I was being removed from the position; he didn't even tell me himself. When I met with him and asked 'why,' he said it was the 'vibe,' that's it. Nothing concrete. And no coaching from him throughout that year. I was eager to learn, and he could have helped"* (Anonymous).

I find that people often neglect deliberately thinking about sponsors and instead rely upon sponsors to step into that role unasked. I believe this is a leadership oversight. When I felt ready for my first corporate board, I approached my mentor, Hugh Rice, to advocate on my behalf. Two years later, the right opportunity presented itself, and Hugh threw my hat in the ring and advocated for me. Proactively seeking out those who have the desire and ability to advocate for you is important.

Activity:

- Think about how you would like your career to unfold over the next three-to-ten years.
- Now think about your criteria for a sponsor, such as having political clout in the organization, having knowledge of you and your skills, or having access to staffing high profile projects.
- Brainstorm a list of possible sponsors within your organization who meet these criteria. Then assess these potential sponsors against the criteria and choose one to approach.
- Write down how you will approach them, and remember to be specific on the "why" and what you hope to gain from the relationship.

Coaches

In 2021, I was struggling to balance the various roles in my life. As mentioned earlier, for 365 days I recorded every hour of time in various categories to help me understand whether I was truly voting for my priorities with my time. At the end of 2021, I knew how much time I spent with my children, my partner, working, learning, exercising—all of it! During my personal retreat to reflect upon this information, I remember writing in my journal, "Ask Edgar for help," because I admired Edgar Schein and knew he juggled many of these same roles—father, consultant, academic, and author. At first, I thought, "He's probably too busy," but it never hurts to ask. Edgar said yes, he would meet with me for 30 minutes. At the end of that meeting, he offered to coach me, and having Edgar as my coach for a year was such an incredible gift.

Coaches differ from mentors and sponsors in that they are typically paid to provide you with professional advice. Someone said to me the other day that they were surprised I have a coach. "Why?" I responded. "Even coaches need coaches!" Each of my coaches has made an incredible difference in my life, and for that I am grateful. I encourage the leaders I work with to think of themselves as elite athletes, to train to show up each day as the best version of themselves as great leaders and to be coached.

> Several days after moving into a new apartment in Melbourne, Australia, in 2023, I met a fellow American in the lobby named David, who invited me for drinks with him and his wife. I asked, "So David, what do you do?" David responded that he was a coach. He would have left it at that, but his wife chimed in and said, "Actually, David is the co-founder of the International Coaching Federation," to which I responded, "You are THAT David Goldsmith?!" (International Coaching Federation, n.d.).
>
> Soon afterward, I spoke with David during a visit to the Greatness Podcast about the value of external coaches. *"Who is going to coach you fearlessly and be able to say, 'Hey, Gretchen, I think your idea here isn't going to work', or 'You said you were going to do this for the last three sessions, and you haven't done it.'? An internal person may not feel they have the safety to be able to have that conversation. I think the outside person also brings a range of perspectives from all the people they're working with, as they're often working with a variety of leaders from a variety of organizations."*
>
> (David Goldsmith, speaking to Gagel, 2023 Greatness Podcast)

Coaches provide expertise to you that is typically organized around a coaching framework. One coaching framework I utilize is the GROW Model published by Sir John Whitmore in his 1992 book *Coaching for Performance* (Whitmore, 2017).

G – Goal – What do you want to achieve with your coach? What does success look like for you?
R – Reality – What is happening today? Getting in the way? Major concerns?
O – Options – What could you do to change this? What are your options?
W – Wrap-Up – What are the next steps, and when will you take them?

This model helps you establish what it is that you want to accomplish at this point in your leadership journey, and I appreciate that it is grounded in the reality of where you are today. Typically, this framework is informed via feedback from others, and a coach helps you explore the many ways you might build a skill or break a habit to become a more effective leader.

Having a coach is a personal thing. As with mentors and sponsors, you must trust them and feel comfortable candidly sharing information. During my last coaching session with Edgar in January 2023, I let him know that my personal relationship of nine years had just ended, and I started to cry. Edgar said, "It's okay to cry," and he sat with me with my tears. He showed incredible empathy, and I appreciated that. Then, we moved on to our discussion of my next steps in my career.

> *"I think about great coaches like a master carpenter. They have an enormous box of tools. They can repair, build, do anything and they know which tools to deploy. Many good coaches do a fine job supporting people, but it tends to be a similar program for everybody. When you move into greatness, you're able to say, this client needs 15 minutes every day, this client needs an hour every two weeks, this client needs a more directive approach, this client needs a lot more listening and provocative questions. Great coaches achieve far greater results in much less time than a good coach and at the fast pace of change that all the businesspeople face they have less and less time. The more that I can get in, get out, help them with what they need surgically and quickly, the more efficient and valuable it is for them. Most importantly, they have to be able to finish that coaching conversation where the client is clearly in action, knows what to do, that nothing's going to get in the way of them taking action because at the end of the day coaching is really about movement and progress."*
>
> (David Goldsmith, speaking to Gagel, 2023 Greatness Podcast)

Effective strategies for finding a coach include asking people you respect for suggestions and visiting a reputable organization's website, such as the International Coaching Federation (International Coaching Federation, n.d.). Before you go shopping for a coach, make sure you know what you are looking for and why. Be clear about what you want to achieve with a coach. When interviewing coaches, ask each one about their methodology and for references from past clients. You might take a coach out to lunch to see if they are a good fit.

> *"**A good coach can change a game. A great coach can change a life.** A coach who actually changes a life changes the game because he is not only striving for gold medal performances, but he's also striving to help create and mold gold medal people. It's the people that ultimately get you the performances and the results at the end of the day. He believed in me more than I believed in myself, and you need those people because people like me, like you, who are very driven, self-critical, very analytical, often it can be hard for us to see our own successes."*
>
> (Anna Meares, speaking to Gagel, 2020 Greatness Podcast)

Coaches, like mentors and sponsors, are an invaluable (and potentially transformational) resource. A lot of the time, your organization will invest in that resource for you. Few athletes would compete without a coach, and I often wonder why so many leaders would do so. I encourage you to explore utilizing a coach as a supportive member of your personal team.

Activity:

- If you have a coach, make sure you are clear on what you want the purpose and outcomes to be for that relationship.
- If you do not have a coach, reflect upon how a coach might add value to your life.
- Approach your organization about possible available resources to support your having a coach.
- Reach out to your network for coach recommendations.

Conclusion

My good friend Cheryl Campbell, former leader of Xcel Energy's gas operations and a Board member of Pacific Gas & Electric (PG&E) and other organizations, mentioned her "Personal Board of Advisors" during a talk with one of my women's leadership development groups. These people are Cheryl's supporting cast (what I call the "raving fans") that she can turn to for assistance and advice. These people challenge and encourage her, provide advice and support, and are indispensable. I cannot think of one great leader I have interacted with over my career who was not adept at creating and leveraging this supporting cast of mentors, sponsors, and coaches.

Many of these great leaders returned the favor to others. Most people love being helpful and appreciate being able to help those they admire. I think of the time I reached out to Edgar Schein for help, and his offer to coach me—such amazing generosity.

"I think ultimately my hope is, as we reflect upon ourselves and who our best selves are, we think about how we can then pay that forward to the people in our lives, think about the impact we have on others and the extent that we can be a part of other's journeys. I think it comes back to what we were talking about the start of the podcast – go be a positive force for good out in the world."

(Amy Jen Su, speaking to Gagel, 2021 Greatness Podcast)

I believe that everybody needs "raving fans," a supporting cast of people you can count on to help you. To live without these people in your life is to forgo tremendous gifts. You do not need to take on all of this at once! Think about what your most pressing priority is right now—a mentor, a sponsor, or a coach. Be brave and ask. Most people love to help, and they receive just as much as you do from the relationship. To ask for help is not a weakness. It is in fact a sign of strength.

References

Browning, A. (2023). Interview by Gretchen Gagel [Zoom]. 23 October.

Buder, L. (1987). 'Beech-Nut is fined $2 million for sale of fake apple juice,' *The New York Times,* 14 November. Available at: https://www.nytimes.com/1987/11/14/business/beech-nut-is-fined-2-million-for-sale-of-fake-apple-juice.html [Accessed: 31 May 2024].

Clark, D. (2018). 'How women can develop—and promote—their personal brand,' *Harvard Business Review*, 2 March. Available at: https://www.hbsp.harvard.edu/product/H046PA-PDF-ENG [Accessed: 31 May 2024].

Ellis, J. (2023). Interview by Gretchen Gagel [Zoom]. 25 October.

Gagel, G. and Edmondson, A. (2019). *Amy Edmondson discusses her book The Fearless Organization on psychological safety* [Podcast]. 30 October. Available at: https://open.spotify.com/episode/6joytIVhV3XlGFZx6rcodC [Accessed: 28 April 2024].

Gagel, G. and Goldsmith, D. (2023). *David Goldsmith, co-founder of the International Coaching Federation, discusses executive coaching* [Podcast]. 15 December. Available at: https://open.spotify.com/episode/1AktDJ16xXLuhPSr6ov1yG [Accessed: 2 May 2024].

Gagel, G. and Mahlab, B. (2023). *Bobbi Mahlab discusses her global organization Mentor Walks* [Podcast]. 22 September. Available at: https://open.spotify.com/episode/0OKfuX2AbZhhHiXB90Odt7 [Accessed: 2 May 2024].

Gagel, G. and Meares, A. (2020). *Anna Meares discusses leadership and team lessons from being an Olympic athlete* [Podcast]. 17 February. Available at: https://open.spotify.com/episode/3k1RqB8K3aiLVG5lRgkRol [Accessed: 28 April 2024].

Gagel, G. and Su, A. J. (2021). *Amy Jen Su discusses her book The Leader You Want to Be* [Podcast]. 12 November. Available at: https://open.spotify.com/episode/4ePtBXmgN40IG7S7pKiFu6 [Accessed: 1 May 2024].

Gosser, M. (2023). Interview by Gretchen Gagel [Zoom]. 14 December.

Gu, T. (2023). Interview by Gretchen Gagel [Zoom]. 11 November.

Henderson, K. (2023). Interview by Gretchen Gagel [Zoom]. 16 November.

International Coaching Federation. (n.d.). *Empowering the world through coaching.* Available at: https://coachingfederation.org/ [Accessed: 31 May 2024].

'Mentor.' (2024). *Merriam-Webster Online Dictionary.* Available at: https://www
.merriam-webster.com/dictionary/mentor [Accessed: 31 May 2024].

Neuscheler, K. (2023). Interview by Gretchen Gagel [Zoom]. 13 December.

Ramsey, A. (2023). Interview by Gretchen Gagel [Zoom]. 16 November.

'Sponsor.' (2024). *Merriam-Webster Online Dictionary.* Available at: https://www
.merriam-webster.com/dictionary/sponsor [Accessed: 31 May 2024].

Sukkar, J. (2023). Interview by Gretchen Gagel [Zoom]. 21 December.

Whitmore, J. (2017). *Coaching for performance: The principles and practice of coaching
and leadership.* 5th edn. London: Nicholas Brealey Publishing.

Whitney, J. (2023). Interview by Gretchen Gagel [Zoom]. 1 December.

8

Experiencing Allyship

Introduction

One aspect of our support systems that I have started paying a great deal more attention to is the concept of allyship, how we show up to support one another and address behavior that is disrespectful and not inclusive. Allyship is critical—woman to woman, man to woman, white to Black—amongst all stakeholders in the construction industry if we are to be truly inclusive.

> "*My earliest memory is of shopping with my grandmother in a supermarket. We were about to check out and the person in front of us, who looked like a giant at that point in my life, was yelling at this poor young girl that was on the checkout. She'd given him the wrong change and he was really having quite a go at her, and before I could blink my five-foot-tall grandmother had inserted herself between this giant and this poor girl, and she pointed her finger up at him and said, 'How dare you talk to that young woman like that. You apologize.' I don't think this man had ever been told off in his life because he took quite a few seconds to collect himself, and he went a bit flushed in the cheeks, and he grabbed his things and sort of mumbled 'sorry' as he walked out of the store. My grandmother proceeded like nothing had happened. She paid for whatever we were buying and walked out of the supermarket before she realized I was still standing back in the line sort of watching all of this play out. She came back to get me, and I said, 'Grandma, that was so brave'. And she said to me something that I've never forgotten, and I think in so many ways has been the marching beat of my life. She said, 'Honey, if you walk past it, you tell the world it's okay.'*"
>
> (Holly Ransom, speaking to Gagel, 2022 Greatness Podcast)

As you read this chapter, consider the ways in which you have received allyship and support from people and how that made you feel. You might also think about a time when you wish you would have taken more of a stance. Be kind to yourself. This is especially difficult early in our careers, and this chapter may provide you with thoughts on how you might have the courage to take action in the future.

Defining Allyship

Allyship is defined as "the state or condition of being an ally: supportive association with another person or group; *specifically*: such association with the members of a marginalized or mistreated group to which one does not belong" ('Allyship,' 2024). Allyship means speaking up when you see something happening that is marginalizing or disrespecting a person or group of people. It requires that you first be present to the unconscious bias (or worse, explicit racism, bullying, or sexism) that causes these situations to arise. Unfortunately, unconscious bias occurs in small and big ways every day in our industry. We, as formal and informal industry leaders in every imaginable role, have the power to be allies, and all people deserve and benefit from allyship. To be an ally is to stand beside someone and tell them they belong, to stick up for someone.

Allies also push us to do things we might not necessarily believe we can do. "*My allies say, 'I'm going to open this door and I'm going to need for you to walk through,' and that's happened time and time again. More often than not they looked nothing like me whether it was male, it was female, white or Latina. They said, 'Denise, I'm here. I can open this door. Walk through because I know you have the credentials.' Then they had my back when I walk through. This made an enormous difference in my career*" (Denise Burgess, speaking to Gagel, 2023 Greatness Podcast).

> "*There are things we can start doing right away and you just gave a perfect example which is asking ourselves, how are we amplifying the voices of women? When you're in that meeting and you just watch that woman's encounter where she brought up an idea and it didn't get a reaction but then later, it could be in the same meeting or down the road, another human brings up the same idea, but she's no longer associated with it or getting credit for it. Amplification is so simple. It's just me saying your name and repeating the great idea. It would be me saying, 'Gretchen, you had that idea two weeks ago, that was a great idea, Gretchen. I want to honor what you just said.' There's something really powerful about naming the person who just had voice around it.*"
>
> (Carrie Arnold, speaking to Gagel, 2022 Greatness Podcast)

There is a crazy YouTube video out there that a professor shared with us during my PhD program (Sivers, 2010). It begins with a scene of a man at a concert dancing in a field by himself. Eventually, another person (deemed the "first follower") joins him. In a few minutes, many people have joined them, and suddenly there is an entire crowd dancing. The video makes the point that while the leader is important, it's the first follower who is critical, in that this person demonstrates that the leader is not crazy! I think of allyship this way. You may not feel that you have the power to stand up for what is right, but you definitely have the power to add your voice to those who do.

Activity:

- Is allyship a concept you consider during your leadership reflections?
- If not, how might you be more present to the concept of allyship?
- Think of examples you have witnessed of great allyship. What made that allyship possible? How might you as a leader encourage more allyship?

Woman-to-Woman Allyship

When I joined a certain company, I was really excited about a woman I thought would be my mentor and ally. No, that was not going to happen. This woman was not interested in explicitly helping me. She did not harm me; she just did not go out of her way to help me or have any interest in mentoring me—probably because she had never experienced woman-to-woman allyship herself. It turned out to be a great experience for me because I promised myself I would never be that woman, that I would be an ally to as many people (and specifically women) as possible.

> *"There is a special place in hell for women who don't help other women. and I believe this to be true of anyone who doesn't help someone else."*
>
> (Albright, 2006)

I was 40 years old before I experienced woman-to-woman allyship, and it began with my dear friend and mentor, Mary Sissel. Mary was Chair of the Board of The Women's Foundation of Colorado when I was hired in 2005, and stood up for me as an ally when a few members of the Board felt I should not be hired due to my lack of nonprofit experience. One of the Board members actually took me out for coffee about three months after I had started and apologized for not supporting my selection, saying I was exactly the right person for the role—good for him! Joy Johnson, who stepped in as Chair of the Board next, was also an ally. As I began my role, Joy called me every week and asked me, "What can I do to help?"

When I became CEO/President of WFCO, my friend and mentor Hugh Rice suggested I seek out Barbara Grogan, a pillar of the Denver community and a former general contractor. Soon after joining WFCO, I approached Barb at a large breakfast gathering where she was being honored, handed her my card, explained that Hugh had asked me to seek her out, and said I would love to meet with her. Barb called me the next day and I was shocked. Here was a person who was on the go, first female on the Kansas City Region Federal Reserve, first female Chair of the Board of the Denver Metro Chamber of Commerce, recent Chair of the Board of Metro State University—the list goes on and on. And she called me! Not only did she call me, but she also helped me, she became my ally, and we became the best of friends.

The love and support I felt were amazing. I knew these women had my back. They were allies. My time as CEO/President of WFCO was an incredible gift to me—not just because of Mary, Joy, and Barb, but also because I had the opportunity to interact with hundreds of fabulous women, many of whom took me under their wing and showed me the ropes in the philanthropic community. They also taught me what woman-to-woman allyship looked like, which was something I had never experienced.

When I stepped down from my role at WFCO, my team threw a surprise party for me, and all my wonderful mentors and allies were there. When I started at WFCO, I set a goal to truly be a member of the Denver community. At the party, then-Mayor John Hickenlooper proclaimed a day for me, and Governor Bill Ritter and his wife Jeannie were there as well. As I looked out at the women and men standing there who had supported me as allies, I was overcome with emotion. Literally, for the only time in my life, I could not make any remarks.

I believe that had I stayed in the construction industry, it might have been another decade or more before I experienced that type of woman-to-woman allyship. When I returned to the construction industry in 2013, I realized what a gift I had been given: the gift of a strong network of women to support me as allies. That had not been possible when I left the construction industry because there just were not enough women. Barb Grogan and I host a dinner party every year for women executives in the construction industry (a wonderful opportunity to build comradery and allyship), and I have now started this tradition in Sydney, Australia, and my hometown of Kansas City. Men often ask me what we say about them at these events. Nothing! We are not here to talk about you (even though we love you!). We are here to create bonds, networks, and allies—to build community.

> *"The most important thing and I think also the hardest thing is that women have to be in community with other women who share their context. If you're the CEO of an organization, you need to be in community with other women running organizations at your level and it's hard to do that because our calendars are so stretched. We can't even find room for our coach to get some coaching, but being in community is essential. It doesn't happen by accident. You don't just stumble onto it. You don't just show up and, oh, here's my tribe. It's intentionality. It is around connecting, showing initiative, asking to be part of it. But it's also keeping an eye out for others and creating space for voice."*
>
> (Carrie Arnold, speaking to Gagel, 2022 Greatness Podcast)

If you are a woman who has never been exposed to women helping you, it might be difficult to trust a woman who reaches out to you to help. When I returned to the construction industry, one of the first women I was introduced to was Teresa Magnus, then with Southern Company. Teresa is an amazingly talented woman, a CPA, and an attorney, and I have great admiration for her. We would catch up at conferences, and I would describe the Teresa I knew then as "cautiously cordial."

I did not think much of it until Teresa and co-author Kathryn Ely wrote an incredible field guide for working on mental health on construction projects titled *Building Mental Wellness: Your Blueprint to Thrive* (Ely and Magnus, 2021). I reached out to Teresa soon afterwards and asked her how I could help promote the book. I invited her onto my podcast and introduced her to people who might want to pilot the program, and I did this for no reason other than to help a woman I saw raising her hand and working hard to fix an issue in our industry. I believe it was then that Teresa began to understand my genuine desire for friendship and camaraderie, and we remain good friends today. I asked Teresa if I could share this story because I realized she had probably been in few positions to have a woman ask her what she could do to help her, to have a woman as an ally. That is the gift I received at WFCO, because if you do not experience woman-to-woman allyship, you might not understand it.

> *"It is very important for these women in this room to go out of this room and be advocates for other women, to make sure that they are doing things to try to lift other women up in their organization or to actively be a mentor to somebody and pay it forward, to focus upon trying to elevate other women. That is something that I'm very passionate about and I spend a great deal of time trying to do."*
>
> (Erica Jones, speaking to Gagel, 2024 Greatness Podcast)

Throughout my interviews for this book, women spoke of women actually being pitted against one another, or viewing each other as "the competition" because of the few positions available to women—especially at the top. *"Women-to-women allyship is hard. Society tends to pit women against each other, making it a competition and thinking there is only one seat at the table versus having four chairs at the table. When I was new to construction, women sought me out to support me"* (Gosser, 2023).

Women supporting women is a critical component of allyship for our industry. We need to approach this industry with a growth mindset, understanding that there is room for all women. If we support one another, we create more success and more inclusion. It is a win-win.

> **Activity:**
>
> - If you are a woman, think of a time when perhaps you did not extend or receive allyship from another woman. What could have been done differently?
> - How does the culture of your team or organization support (or not support) the concept that there are many seats available to women at the table? What might you do differently to support this concept?
> - How can you, as a person of any gender, do more to support woman-to-woman allyship?

Male Allies of Women

I interviewed my first FMI supervisor, Lou Bainbridge, for this book because I wanted to ensure that my recollection of this situation was correct. When I first joined FMI in 1994 (which was then the largest investment banking/management consulting firm focused exclusively on the construction industry), to my knowledge, I was the second woman hired—and at that time, we employed over 100 consultants and advisors. Early in my career at FMI, I was leading quite a bit of "project partnering" on construction projects and strategy work for construction companies. Some construction companies asked that I not be assigned to work with them on their strategic plan because of my gender (I know, shocking!). However, the more shocking thing to me today was Lou's response back then: "Well, you'll need to find another firm because Gretchen is an important member of our team" (Bainbridge, 2023).

30 years later, I now realize how unusual this allyship was at the time. I asked Lou why he responded this way, and his answer was beautiful in that it connects back to our mutual mentor, Hugh Rice. Lou said that Hugh set the example by focusing on the importance of relationships, respect, and "doing the right thing" (Bainbridge, 2023). Lou also mentioned our mutual friend, the late Peter Nosler, co-founder of DPR Construction, as someone who had taught him that we need

to stand up to things we believe are not right. *"It's my responsibility to advocate for women in our industry because I am a white male. If you are silent you are contributing in one way or another"* (Bainbridge, 2023). Amen.

> *"Once we create that community with women, we also have to let men into that community because they are our allies, and they are eager to amplify our voices."*
> (Carrie Arnold, speaking to Gagel, 2022 Greatness Podcast)

The topic of allyship came up frequently during my interviews for the book as these women and industry leaders shared their experiences with allyship:

> **Kerri Smith Petrillo, Chief Talent and Strategy Officer, Baker Construction:** *"The first superintendent I worked with as a project engineer was an ally. I worked alongside him all day, outworked everyone, and asked questions. He had my back. He supported me, treated me differently because he saw my dedication, and recognized my eagerness to learn – not because I as a woman."* (Petrillo, 2023).

> **Laura Stagner, FIAI, DBIA, PMP, Retired Assistant Commissioner, Office of Project Delivery, GSA:** *"Yes, I've been in meetings where I stated an idea and two minutes later a man repeated my same idea, and the man got praised for having had such a great idea. I have also had the opposite experience – someone in a meeting says, about some topic about which I am an expert, 'you should consult (man) about this', and another man will speak up and say, 'no, you want to speak to Laura about that, it's her area of expertise.'"* (Stagner, 2024).

> **Josephine Sukkar, Founder and Principal, Buildcorp**: *"My biggest enemy was my own self talk. Men lifted me up, convinced me to do things"* (Sukkar, 2023).

Male construction industry leaders also discussed the important of allyship during their interviews:

> **Tom Reilly, President, Turner Construction Company:** *"I encourage people to look back at how they advanced in their careers. I ask them to reflect on what it was like for them versus people with a different background. I find that this helps people to become advocates for others."* (Reilly, 2023).

> **Steve Dora, Senior Manager Facility Projects, Toyota North America:** *"I pushed her out of her comfort zone by asking her to do a weekly presentation,*

to take ownership and be the voice of safety. I helped build her confidence by doing a quick review of her slides ahead of time, saying, 'You've got this' and giving her pep talks" (Dora, 2023).

Brian Gallagher, Vice President Corporate Development, Graycor Companies: *"I encourage men in our industry to be open-minded. Don't rely on bias or stereotypes about a female's ability to work in the industry. Think about what you can do to help, what can you learn from women, how you can help be an advocate for change"* (Gallagher, 2023).

Dominic Rotunno, Senior Director, Capital, Engineering, Equipment, and MRO, Brystol Myers Squibb: *"Diversity is important. Women, some other minority, everyone has a story, has value to bring to the table. Be inquisitive, ask them for their opinion, facilitate what they can bring to the table"* (Rotunno, 2023).

I have tremendous admiration for people whose life work is helping us understand and overcome bias and what steps we can take to become allies. For over 15 years, Jeffery Tobias Halter, Founder of YWomen, has worked to help men understand how to support women and other underrepresented communities in the workplace as allies. *"I find there's four barriers to men not becoming advocates. First lack of empathy, thinking, 'I don't believe men and women are really having that big of a different experience.' Apathy/Failure to internalize the critical need for change now – 'it's 2024, can't we all get along?' Third is lack of accountability - it doesn't affect me or my paycheck and no one holds me accountable for diverse slates, diverse panels. The last one is fear. Men are still scared that we will say or do the wrong thing and so they CHOOSE to do nothing."* (Jeffery Tobias Halter, speaking to Gagel, 2023 Greatness Podcast).

I was walking at a job site where the female project manager shared the story of a recent incident when a member of the framing crew had whistled at her. When she shared this with one of the male superintendents, he immediately gathered the entire framing crew, stated that she was their boss and was to be treated with respect, and called the framing company's owner, who supported his actions and asked that he be informed of any future transgressions. This exemplifies zero tolerance for disrespectful behavior and man-to-woman allyship.

I encourage men not to be afraid to point out a problem in our industry, even if you do not have the solution. In September 2018, two weeks after arriving in

Australia, I presented a paper on energy company/contractor alliances in the US at the APGA conference in Darwin. At that conference, Steve Davies, CEO of APGA, bravely stated to the audience that he was embarrassed there was not a woman on stage during his senior leader panel. Steve said he did not know what to do to bring more women into the industry, but that we needed to work on it.

I took Steve out for coffee a few weeks later and described the American Gas Association's (AGA) women's leadership program, whom I had spoken to many times. I suggested to Steve that we replicate the program with AGA's approval; Steve ran the idea past his board, and five years later, we had completed nine cohorts of the program with nearly 200 participants. During one outcome survey, 90% of the participants reported an increase in confidence, and quotes like these exemplify the program:

> *"After 10 years in a stalled career I've had two promotions in two years. This program has transformed my life."*

> *"It made me more empathetic to people in my business – that we all suffer from insecurities. I have also found that people come to me more for guidance and support. I feel more confident in my role."*

> *"The best value of a course is when you can get at least one action completed that you set for yourself. I was able to find a female mentor, have tough conversations with my superiors at work in terms of feedback and understand that I can only control my actions and for that, I will be extremely grateful. I hope that I continue to learn from the experts and amazing women that I met on this journey and possibly inspire others in the future."*

This is only one of many strategies needed to improve the inclusivity of the pipeline industry, but it is a step in the right direction. Steve did not know what the solution was, but he had the courage to point out the problem and ask for help solving it.

> *"Standing up for what's right and being brave enough to do that is critical because I also know that it's really difficult as a man to be the advocate for the women in the room. There's a chance that you are also placing yourself outside of the good old boy's club if you're doing that, and so they have to be brave enough to stand up and create inclusion, to actually think about it, to change some of their verbiage that they are using and be intentional about that."*
>
> (Erica Jones, speaking to Gagel, 2024 Greatness Podcast)

When I became pregnant soon after joining Lou's team at FMI (a job that required traveling at least four days a week), Lou said, "We'll figure this out," and "This is just a new piece of information;" it did not feel like a big deal to him, even though it was a really big deal to me (Bainbridge, 2023). I was nervous about how I could continue to do my job and be a mom. Lou said that I had this "can do" attitude, that nothing was going to phase me, and that it was not a negative. That is male allyship of women in the construction industry.

Activity:

- Reflect on the times that you have either observed or experienced male allyship as a woman. What did that look like? What made it possible?
- If you are a man, what examples can you think of where you or another male have been an advocate to a woman? How might you promote more of this allyship?
- Perhaps you've looked past a specific behavior at some point. What would you do differently today? What might give you the courage to act?

Conclusion

Perhaps you are not in a position within your team or organization to dictate policy, but you can lead, and you can certainly be the first follower when someone else leads. You can be present to and celebrate allyship by others, and you can be an ally yourself. Allyship can happen in big and small ways. When someone speaks over a woman in a meeting, you as a person of any gender are able to ask her to finish her thought, and perhaps privately point out the behavior later, as many times, people do not realize they are speaking over people. When someone is suggesting only male candidates for a position, you can suggest diverse candidates.

"Sixty-three percent of women experience a microaggression on a regular basis. When you're working in a group, just keep a tally sheet of how many times a woman's voice is interrupted or ignored, and the number will surprise you."

(Jeffery Halter, speaking to Gagel, 2023 Greatness Podcast)

Allyship is each of us valuing all human beings for their contributions to society and our industry. It is helping each person feel like they belong and advocating for them. We all have the power to be allies, and I believe there are several elements that give us the courage to do so: (1) our core values and moral compass; (2) our willingness to be cast aside from a group for being an ally to another; and (3) our belief in humankind and that others will join us. I encourage you all to be brave and support each other as allies.

> *"Find the women who deserved to be advocated for and advocate for them. See people as people and find people that are worth advocating for."*
>
> (Magnus, 2023)

References

Albright, M. K. (2006). 'Madeleine K. Albright | Quotes | Quotable quote,' *Goodreads*. Available at: https://www.goodreads.com/quotes/14328-there-is-a-special-place-in-hell-for-women-who [Accessed: 10 June 2024].

'Allyship.' (2024). *Merriam-Webster Online Dictionary*. Available at: https://www.merriam-webster.com/dictionary/allyship [Accessed: 10 June 2024].

Bainbridge, L. (2023). Interview by Gretchen Gagel [Zoom]. 21 November.

Dora, S. (2023). Interview by Gretchen Gagel [Zoom]. 23 October.

Ely, K. and Magnus, T. (2021). *Building mental wellness: Your blueprint to thrive*. Atlanta: BDI Publishers.

Gagel, G. and Arnold, C. (2022). *Carrie Arnold discusses her book Silenced and Sidelined—How Women Leaders Find Their Voices and Break Barriers* [Podcast]. 19 August. Available at: https://open.spotify.com/episode/0QnMQ73dUBxPBJNkOQEM3V [Accessed: 1 May 2024].

Gagel, G. and Burgess, D. (2023). *Denise Burgess discusses her career as an African American woman leader in the construction industry* [Podcast]. 29 December. Available at: https://open.spotify.com/episode/77UCAi1FXUKeMOS6yfes8z [Accessed: 2 May 2024].

Gagel, G. and Halter, J. T. (2023). *Jeffery Tobias Halter discusses how we create inclusive organizations* [Podcast]. 17 November. Available at: https://open.spotify.com/episode/5HcF8ZcTU01guNNUsb8UI2 [Accessed: 2 May 2024].

Gagel, G. and Jones, E. (2024). *Erica Jones discusses women's leadership in construction* [Podcast]. 23 February. Available at: https://open.spotify.com/episode/3RQth0SyIU9nWVsm4RSqKg [Accessed: 2 May 2024].

Gagel, G. and Ransom, H. (2022). *Holly Ransom discusses her book The Leading Edge* [Podcast]. 30 September. Available at: https://open.spotify.com/episode/0M5G6yRwqvbBlKa302wmKb [Accessed: 1 May 2024].

Gallagher, B. (2023). Interview by Gretchen Gagel [Zoom]. 25 October.

Gosser, M. (2023). Interview by Gretchen Gagel [Zoom]. 14 December.

Magnus, T. (2023). Interview by Gretchen Gagel [Zoom]. 9 November.

Petrillo, K. S. (2023). Interview by Gretchen Gagel [Zoom]. 20 November.

Reilly, T. (2023). Interview by Gretchen Gagel [Zoom]. 25 October.

Rotunno, D. (2023). Interview by Gretchen Gagel [Zoom]. 24 October.

Sivers, D. (2010). *First follower: Leadership lessons from dancing guy*. Available at: https://www.youtube.com/watch?v=fW8amMCVAJQ [Accessed: 10 June 2024].

Stagner, L. (2024). Interview by Gretchen Gagel [Zoom]. 25 October.

Sukkar, J. (2023). Interview by Gretchen Gagel [Zoom]. 21 December.

9

Building Powerful Networks

Introduction

Soon after I moved to Australia, a friend mentioned that she had met a woman in the US who was a senior leader with a global engineering firm based in Australia, and asked if I would like to meet her. "Of course!" We had a virtual catch-up, and this new friend offered to introduce me to her company's leaders in Australia, including the Chair of the Board. "Of course, that sounds great!" Three years later, I was offered a position on one of their Board Committees, and it all started with me saying "yes" to a meeting without any idea of where it would lead. Your networks are another critical source of support and opportunity for you as a leader. Not only is building networks good for you professionally, but the World Happiness Report also provides evidence that positive social connections are one key factor in personal and community happiness (Wellbeing Research Centre, University of Oxford, 2024).

Networking is defined as "the action or process of interacting with others to exchange information and develop professional or social contacts" ('Networking,' 2024). For many of us, building networks requires a great deal of energy. Even though I am a natural connector, moving to a new country at the age of 54 required seemingly endless networking to reestablish myself and start my company. It was intimidating, and at times exhausting, but vital to the success of my new business. Introverts can be excellent networkers, but they must understand that time with people also creates a need for time alone to recharge their batteries. You might feel intimidated attending a conference where people are more tenured than you, or do not look like you, or speak a different language than your first language. Some people may actually fear meeting other people.

> *"Networking is really relationship building. I think of it in two directions. You can build your relationship or your network out, get longer and broader and meet more people; or network deeper which is to take relationships that you have and deepen those relationships. I would rather do the work ahead of time, would rather build my network when I don't have an ask. Then when the time comes that you have a need you will already have the foundation of your relationship there. Call it whatever you need to call it – if you need to call it 'going for coffee,' call it 'going for coffee,' because most people know how to do that. I am a huge fan of quick wins. If it is February and I'm going to go for coffee once with somebody I haven't gone to coffee with before or haven't in years, you get to count that as a win. Don't set the bar so high that you are set up to fail, set the bar low."*
>
> (Deborah Grayson Riegel, speaking to Gagel, 2024 Greatness Podcast)

If you've read *The Tipping Point: How Little Things Can Make a Big Difference* by Malcolm Gladwell, you are familiar with "connectors," people who naturally enjoy networking (Gladwell, 2002). Connectors also enjoy linking people to other people in ways that are helpful to others. I believe the energy you dedicate to building networks is important because of the joy of finding and being in relationships with like-minded individuals. Your networks provide support, encouragement, and opportunity.

As you read this chapter, think about all of the positive ways networks have contributed to your happiness and success, and consider new ways you might invest in your networks both within and outside of your industry.

Your Industry Networks

One thing I love about the construction industry is that most of us genuinely enjoy spending time with one another, and your industry associations are a great starting point for networking. I feel fortunate that the first person I met professionally in Australia was Mark Bumpstead, then-CEO of Quanta Services Australia, who encouraged me to join APGA. This was a great starting point for building my network in a new country, and many of my best friends in Australia have come from my involvement in this association.

In the United States, we have general construction industry associations with local chapters in most markets such as the Associated General Contractors (AGC); specific industry associations by type of industry such as the American Gas Association; associations for different types of contractors such as the Sheet Metal and Air Conditioning Contractor Association of America; and associations for women in our industry such as the National Association of Women in Construction (NAWIC). So many choices!

> I encourage you to become involved in at least one association for the following reasons:
>
> 1) These associations are hard at work making individuals, companies, and the industry better, and they can use your help.
> 2) These associations provide you the opportunity to interact with and learn from like-minded individuals.
> 3) These associations also provide you the opportunity to be exposed to future employment opportunities.

Women are often outnumbered in these associations, and that can make joining and attending association events intimidating. If you are new to industry networking, I suggest you try attending one event with one association as a starting point. If networking is hard for you, try attending an association event or conference with a small goal, such as making one new friend at a meal or cocktail party. Find someone else sitting alone at lunch, ask if you can join them, and say "Hi, I'm XXXXX with XXXX, nice to meet you." Think of one or two great open-ended prompts ahead of time like, "Tell me about what you do," or "Tell me about your organization," or "What session has been your favorite so far, and why?"

Not all industry networking is accomplished at association events. In 2023, I decided to buy a home in Kansas City (the hometown I had left in 1982) because my grown children lived there. While I already knew a handful of construction industry friends in Kansas City such as former Black & Veatch CEO Steve Edwards, I decided to make a conscious effort to "make new friends" because the construction industry is my tribe and an important priority for me. I began by transferring my AGC membership from Denver, Colorado (where I had lived for 27 years) to The Builders, a chapter of the AGC, in Kansas City. I made an appointment with Angela Crawford, Membership and Marketing Vice President, and explained my experience in the industry. I then asked Angela for suggestions of who I should meet in the industry there. It turned out that one of their former board chairs lived in my new building, so not only did I meet someone in the industry, but I also met a new neighbor!

Another easy step was to research a list of the largest construction and engineering firms in Kansas City and their senior leadership teams. I sat down one afternoon and reached out to each person on these teams via LinkedIn with a short note that said, "I am moving back to my hometown of Kansas City after 40 years of advising in the construction industry in the US and Australia and would love to connect." I met with the people who responded, and at the end of every meeting, each person offered to introduce me to more people. A few months later, I had the joy of hosting a dinner for 18 women executives in the construction industry in Kansas City, my new tribe. Yes, it took some effort, but I am thrilled to have new friends and supporters in the construction industry.

When I asked Avery Bang, former CEO of Bridges to Prosperity and Partner at the Mulago Foundation, "Who are the women you most admire in the construction industry?" she began naming the women construction leaders I regularly convene with (including my friend Barb Grogan) for a social dinner in Denver, Colorado (Bang, 2023). I asked Avery if there were any other women in the industry she admired, and she could not think of one. Had I not invited Avery to these dinners, she would have had an extremely limited network of women in the construction industry. Out of curiosity, I also asked Avery how many of the primary contacts of the 70+ construction and engineering corporate partners she interacted with while CEO of Bridges to Prosperity were women. Her answer, none, astounded me.

> Women-centric networking events like the On the Rise program for women in construction that Erica Jones, Vice President of Marketing for Cerris (formerly MMC Corp), founded in Kansas City are a great opportunity for networking. *"These women walk away feeling so inspired. They've increased their own network and several of them have created friendships that last beyond that event. People come back year after year because they are walking away feeling inspired and connected to all of these other people in our community. I am a female in a fairly traditional female role, and I recognize that it requires even more bravery for those women that are in more male traditional roles. Sometimes they want to be disassociated with things like the women's professional network or women in construction week or NAWIC because they don't want to be seen as the female or revered because they are a female in construction. I say to them, it is okay that we are different. It's okay that we bring something different to the table. We should be embracing that and celebrating that. That's what makes us all better. We shouldn't be saying we don't want to acknowledge the fact that we're female because we don't want to cause any issues or be seen as different because that is what makes every organization better, a blend of diverse backgrounds and mindsets and communication skills."*
>
> (Erica Jones, speaking to Gagel, 2024 Greatness Podcast)

Here are some ways to start small with networking in the construction industry:

Industry events:

- Attend one industry event and make three new friends. Make sure you either exchange business cards (or, as many people do not carry cards anymore, I will ask if I can take a picture of their nametag, and then later connect with them on LinkedIn).

- Send a follow-up email, and possibly schedule a follow-up coffee.

Company networking:

- Think of someone you've been introduced to via a recent meeting, etc. within your company. Maybe they are on a different team or in a different department that you want to learn more about.
- Reach out to them and say something like, "Hey, I enjoyed your presentation (or your comment, or your insights), would you like to have coffee sometime?"
- Think of someone you have met who works for a similar company in the industry. Consider taking them out to coffee.

I believe it is important to develop an industry networking plan that meets your individual needs for learning and social interaction and considers your natural affinity and enjoyment of networking. Industry networks can be a great source of support and opportunity. Creating an industry networking plan that leverages your strengths and does not feel like a burden is important as you develop as an industry leader.

Activity:

- Think about the sources of industry networking you currently enjoy. What are the benefits to you?
- Is industry networking something you should expand? If so, in what way?
- What association event might you attend? What three women or men might you invite to lunch to interact with as a network?

Networking Outside of Your Industry

When I became CEO/President of WFCO, networking with women was a completely foreign concept to me since I had spent the past 20 years in manufacturing and construction. My wonderful female mentors nominated me for the International Women's Forum, and I joined in 2008. What a great decision this was, and one that underscores the value of networking with people outside of your industry. This group of approximately 8,000 women in 30+ countries has provided me with friends around the globe. Even more importantly, when I moved to Australia, I had a built-in network of 150 accomplished women who took me under their wing and helped me make connections as I started my new company and life.

Networking outside of your industry is also important for the diverse perspectives it brings to you as a leader. I had the good fortune during my PhD studies to take an entire course on scenario planning with Dr. Thomas Chermack, one of the world's leading experts on the topic (Chermack, 2011). Tom is quite a student of Pierre Wack, who created many of the important concepts of scenario planning while working at Shell (Chermack, 2017). One of the important elements of scenario planning stressed by both Pierre and Tom is the need for leaders to collect information from a variety of sources, including less obvious sources that exist outside of your industry. This information ensures that you, as a leader, do not suffer from tunnel vision and are armed with a wide array of data that informs the strategies your organization will need to deploy in order to sustain success. If you are a doctor who only spends time with doctors, or a solar engineer who only spends time with solar engineers, you are limiting your access to broader information and ideas that will enhance your critical thinking.

My friend Lisa Barron, a Melbourne-based fashion designer, is a great example of the importance of networking outside of your industry. Lisa was one of the women I met with every Wednesday during our astounding 262 days of lockdown in Melbourne, Australia, from March 2020 to October 2021 (Zhuang, 2021). During one of our weekly virtual gatherings, Lisa was discussing her new clothing collection and the limited options she had to promote and sell these clothes, given the fact that her store was closed. I suggested a virtual runway show, and I actually ended up emceeing two virtual runway shows for her over the next year—the second with over 100 people online, and both with tremendous success. If Lisa had only been spending time with people in the fashion industry, she might never have been exposed to this idea.

Here are some ideas of how to broaden your network outside of your industry:

- Network-creating events (such as the Mentor Walks I spoke of earlier) are a great way to network with women outside of your industry.
- Women's organizations are another great way to meet women outside of your industry, and these organizations exist in nearly every city. For example, most United Way organizations and community foundations have some type of women's network.
- The local Chamber of Commerce and Rotary Club are two additional good starting points for networking outside of your industry.
- Nonprofit organizations and events are another great way to support a cause you believe in, give back to your community, and network with people outside of your industry.

I strongly encourage you to give some thought to developing a strategy for networking outside of your industry. Even if it is one nonprofit event or mentoring event a year and you meet three people, these three people have the potential to increase your social connections and expand your thinking.

Activity:
• Think about your current business and social networks outside of your industry. What are the benefits you receive from these networks? • Take a few minutes and brainstorm more ways you could expand your network outside of your industry. • What one first step might you take?

Conclusion

Your ability to develop networks in a way that is authentic to you is critical to your success as a leader. These networks are a source of power, advice, job opportunity, and support. It may require a bit of energy to develop these networks, but I believe the return is worth it.

> *"Networking is critical for career and personal growth. I ask leaders, who are your intelligence bank, your board of advisors, your personal marketing machine who understand what it is that you're trying to achieve over the next twelve months, three years? As I describe in my book,* It's Who You Know, *there are essentially 12 key people you need in these four groups: 1) Promoters: They are your marketing machine, they believe in you. they see more in you than you see in yourself; 2) The Pit Crew: Their role is to keep you in the moment, to care for you, to make sure that you are getting that balance that we all need physically, mentally, and spiritually; 3) Teachers: These are the people who are committed to helping you know more and the reason these people are so important is because the competitive advantage that we all have right now is what we think; 4) Butt Kickers: These are the people that are calling you on your BS, that are mentoring you and checking in every single day, that are accelerating those goals."*
>
> (Janine Garner, speaking to Gagel, 2022 Greatness Podcast)

Ensuring you have a diverse network of people to support you is critical.

References

Bang, A. (2023). Interview by Gretchen Gagel [Zoom]. 11 December.

Chermack, T. J. (2011). *Scenario planning in organizations: How to create, use, and assess scenarios*. Oakland: Berrett-Koehler Publishers.

Chermack, T. J. (2017). *Foundations of scenario planning: The story of Pierre Wack*. New York: Routledge.

Gagel, G. and Garner, J. (2022). *Janine Garner discusses her book Be Brilliant* [Podcast]. 14 October. Available at: https://open.spotify.com/episode/ 2jTwgPFgU2jmhi9kG6WNTW [Accessed: 1 May 2024].

Gagel, G. and Grayson Riegel, D. (2024). *Deborah Grayson Riegel discusses networking and women's relationship with failure* [Podcast]. Available at: https://open.spotify .com/episode/1eSA809JUvn8g9YfgSlkgt [Accessed: 15 July 2024].

Gagel, G. and Jones, E. (2024). *Erica Jones discusses women's leadership in construction* [Podcast]. 23 February. Available at: https://open.spotify.com/episode/ 3RQth0SyIU9nWVsm4RSqKg [Accessed: 2 May 2024].

Gladwell, M. (2002). *The tipping point: How little things can make a big difference*. New York: Back Bay Books.

'Networking.' (2024). *Merriam-Webster Online Dictionary*. Available at: https://www .merriam-webster.com/dictionary/networking [Accessed: 10 June 2024].

Wellbeing Research Centre, University of Oxford. (2024). *World happiness report 2024*. Available at: https://worldhappiness.report/about/ [Accessed: 10 June 2024].

Zhuang, Y. (2021). 'Melbourne, after 262 days in lockdown, celebrates a reopening,' *The New York Times,* 22 October. Available at: https://www.nytimes.com/2021/10/ 22/world/australia/melbourne-covid-lockdown-reopening.html [Accessed: 10 June 2024].

Part III

"We" – The Supporting Cast – Conclusion

The raving fan I miss the most is my mother. No matter what crazy career decision I made, she was there cheering me on. When I was having a crisis about my ability to be a mother and have a career, she wrote me a letter that I treasure.

> *"You should feel good about yourself as a mother and not sell yourself short. Remember that you won't be a mother like me, like Dan's mother, or like anyone's image of the perfect mother. You just need to be yourself and raise your children in the best way you know how. Looking at Regan and Holden you certainly shouldn't feel you are less than adequate. You are very special and I love you dearly, Mom."*

Thinking strategically about your supporting cast—your raving fans—is a sign of intelligence and thoughtful leadership. Throughout your life and leadership journey I encourage you to be deliberate in thinking about your mentors, sponsors, coaches, allies, and networks. Each has a gift for you if you are open to receiving it, and you have the ability as a leader to fulfill these roles for others.

Part IV

"We" – Building Strong Relationships – Introduction

Years ago, retired Major General Simone Wilkie came to guest lecture for a course he was teaching at the Australian National University (ANU) on leadership. Simone is an incredibly gifted and respected leader who made outstanding contributions as a leader in the Australian Army during her 35-year career, and continues to make her mark as a leader on several boards. Simone explained that early in her career, she treated each relationship with a person as a bridge with explosives and a detonator at the end. If she did not like how the relationship was going, she would just "detonate" that relationship. As her career progressed, Simone realized that this was not an effective leadership strategy, as leaders must build relationships with people they like and trust, people they do not like, and even people they do not trust. You cannot just walk away because it is via your relationships with people that you lead and influence. I will say that in very few instances, you should walk away.

Early leadership theory was often focused upon the traits of a leader, such as charisma, confidence, and outgoingness (Burns, 1978). However, more recent leadership research has placed greater emphasis on the relationship and interaction between the leader and those they are leading and influencing (Schein and Schein, 2023). Our ability to build relationships with all stakeholders (like those we lead and coach, peers, supervisors, customers, and suppliers) is critical to our success as leaders and the success of these stakeholders.

Building relationships is not a natural skill for many of us. Friend and retired mining executive and Non-Executive Director Jacqui McGill once told the story of being passed over for a couple of promotions early in her career in the mining industry. The company hired a coach for her, and this coach asked Jacqui if she knew the names of any of her team's spouses, or whether they had kids or dogs. Jacqui admitted she did not. Jacqui was passed over for these promotions because although she was very good at accomplishing tasks, she was not as adept at building relationships. She was leading, but her people were not following. The turning

Building Women Leaders: A Blueprint for Women Thriving in Construction,
First Edition. Gretchen Gagel.

point for Jacquie was reading the book *How to Win Friends and Influence People* by Dale Carnegie (2022). Building relationships was not a skill that Jacqui had developed, nor a muscle she had strengthened. Jacqui used the book and her coach to build that skill. She started taking the time to learn about each of the people on her team, to build relationships such that they knew she cared about them as people—not just as someone accomplishing work.

> *"The key to humble leadership is that we have established a whole person-to-whole person relationship. We decided to use the term personize because it's different than personalized. You establish an open trusting bond based on revealing something about yourself or inquiring about the other person. It's something we do all the time with our friends and our family, but it's something that since the 1950s we've been a little bit more reluctant to do in the work environment. We still have a dominant myth about professional distance being the most appropriate way to conduct yourself at work. We don't really believe in that anymore and that's part of our message around humble leadership, is that professional distance is an antiquated idea. It doesn't help you get through complex situations. It may be fine for executing transactions between roles, but professional distance gets in the way of creativity, of innovation, of adaptation in complicated or complex situations. These days at work everything is a complicated and complex situation."*
>
> (Edgar Schein and Peter Schein, speaking to Gagel, 2019 Greatness Podcast)

During my PhD studies, I researched the work of Bernard Bass, who built upon the seminal work of James McGregor Burns to build the theory of transformational leadership (Bass and Riggio, 2005). I initially thought "transformational leadership" referred to the ability to transform teams and organizations, but through my study of his work, I learned that it in fact means the ability to transform individuals, which in turn facilitates the transformation of teams and organizations. One of the key elements of Bass's model of leadership is "individual consideration," taking time as a leader to learn about everyone on the team—including their wants and desires, what motivates them, and their preferred communication style. The time a leader takes to deeply understand each team member, to say, "I see you," "I hear you," and "I value you," is an incredible gift. Edgar and Peter Schein call this "personization," the building of "Level 2" relationships (Schein and Schein, 2023).

> Poet Maya Angelou writes, *"I've learned that people will forget what you said, people will forget what you did, but people will never forget how you made them feel."*
>
> (Angelou, n.d.)

I have spent a great deal of time as a PhD student, professor, coach, consultant, and observer developing my own model of what I believe to be the most important skills for building strong relationships with individuals:

- It starts with being fully present to others, listening and asking great questions to understand.
- As leaders, we must also understand our sources of power and how to adjust our power to a level that is appropriate to that context and enables us to influence the behavior of others.
- Fundamental to this is our emotional intelligence, how we understand our own emotions and our "triggers," and have empathy for the emotions of others.
- As leaders, we must also be adept at building trust and psychological safety, as people do not follow people they do not trust.
- The three "C's," as I call them (communication, collaboration, and conflict resolution), are all critical in our relationships with those we lead and influence.
- Our ability as leaders to clarify our expectations and constructively provide feedback is also a gift to others if done well.

As you explore each of these topics, I encourage you to reflect upon yourself and build on the thinking you explored in the Leading Self section. None of us will be Olympic athletes in every skill. Celebrate the times when you read something and think, "Oh, I'm good at that!" and take note when an idea or story causes you to think, "That is definitely something I could work on."

References

Angelou, M. (n.d.). *'Quotes,'* Goodreads. Available at: https://www.goodreads.com/author/quotes/3503.Maya_Angelou [Accessed: 6 July 2024].

Bass, B. M. and Riggio, R. E. (2005). *Transformational leadership.* New York: Psychology Press.

Burns, J. M. (1978). *Leadership.* New York: Harper & Row Publishers.

Carnegie, D. (2022). *How to win friends and influence people: Updated for the next generation of leaders.* New York: Simon & Schuster.

Gagel, G., Schein, E., and Schein, P. (2019). *Ed & Peter Schein discuss their book Humble Leadership* [Podcast]. 24 June. Available at: https://open.spotify.com/episode/5JueFC2LHdo74Ncz4HqVs5 [Accessed: 28 April 2024].

Schein, E. H. and Schein, P. A. (2023). *Humble leadership: The power of relationships, openness, and trust.* 2nd edn. Oakland: Berrett-Koehler Publishers.

10

Being Present, Listening, and Asking Great Questions

Introduction

When you walked into someone's office in the 1980s, there was no computer on the desk, no cell phone pinging—just you, that person, probably a large wooden desk, and stacks of paper. Yes, someone could be looking at a piece of paper instead of you, but for the most part, offices created space for focused interaction. Then came cubicles, and computers, and cell phones, and Slack, Yammer, Teams—the many ways that people and electronics are vying for your attention. Sometimes, I find that I have so many competing ideas happening in my brain that I am not fully present when interacting with someone, and this upsets me because I know that being present is critical to building strong relationships with those I lead and influence.

To be fully present requires listening and asking great questions, being an "asking leader" versus a "telling leader" (Schein and Schein, 2021). A leader can be intently listening and still remain focused on sharing their views and opinions versus seeking to understand the viewpoints and opinions of others. Being an "asking leader" means authentically probing into what that person is thinking and feeling by asking great questions at the right moment (Schein and Schein, 2021). This is an incredible leadership gift.

> "*If you're a leader adding coaching skills to your tool bag, I think what you get really good at is listening, asking questions, and starting to shift from being directive to being curious and being able to help other people figure things out. You provide good examples, good questions, good thinking, good processes, because if they can figure it out for themselves, they're going to grow, they're going to be able to do it again the next time without you. That's the measure of what you want as a coach or a manager or a leader.*"
>
> (David Goldsmith, speaking to Gagel, 2023 Greatness Podcast)

As you read this chapter, I encourage you to be fully present. Think about yourself in each context described and consider both your strengths and your challenges. Be kind to yourself. Today's busy work environment creates many barriers to perfecting these skills. Remember, you are a work in progress and the point is to learn and improve.

Being Present

My mentors all share one incredible leadership skill. When I sit down to ask one of them for their advice, I feel as though I am the only person in the world. They are not looking at their phone, or at anyone else in the restaurant or office. They are not thinking about other things on their to-do list. Each one is intently focused on me, listening to my situation with empathy, and asking me great questions that expand my thinking on that topic. That is leadership. Being present "... *simply means you're focused and engaged in the here and now, not distracted or mentally absent*" (Raypole, 2020). Focused and engaged: two simple words that reflect what can be challenging but critical to great leadership. *"Being in the present moment, or the 'here and now,' means that we are aware and mindful of what is happening at this very moment. We are not distracted by ruminations on the past or worries about the future but centered in the here and now. All of our attention is focused on the present moment"* (Thum, 2008; Ackerman, 2018).

Being present is an art that, like anything else, takes practice. Meditation and breathing exercises train our brains to be present in the same way that going to the gym trains our bodies. Practicing these techniques, reducing "task switching," and eliminating distractions such as your cell phone are all strategies that contribute to you being able to be a more present, focused, and engaged leader.

> I used to think I was the "queen" of multitasking (or "task switching") and took great pride in my ability to easily switch from task to task throughout the day—teaching, coaching, consulting, writing. Now I know that task switching exacts a cost, and I believe task switching is a significant contributor to our inability to be focused and engaged with those people we lead and influence. We lose 5–10 IQ points when we task switch, which, as friend and author Kristen Hansen says, is about the same as being drunk, stoned, or staying up all night (Hansen, 2018). What leader wants to "show up" to their teammates as that?! When we "task switch," tasks take more time, we make more mistakes, and it can lead to up to a 40% loss in productivity
>
> (Weinschenk, 2012)

Research shows that just having a cell phone on your desk reduces your cognitive ability (Ward *et al.*, 2017). I now try and put mine in my briefcase or in a drawer for most of the day. Occasionally, I will post a note on the wall in my office that says, "Slow down," to help me remember to be present. I also believe part of the challenge of distractions is that we all (including me) are too nice. When someone walks into the job trailer and wants to speak with you, even if you are deep in thought about another topic, you might instantly turn your attention to them. I have learned that it is okay at times to say, "You know what, my head is in a completely different space. Can we talk about this in 30 minutes?"

> *"You've just got to do that by really listening and being present and not thinking you must be the smartest person in the room and have all the answers. I think great leaders know what's important to them and know what's real to them. But they also spend a lot of time knowing what's important and what's real for the people that they lead and that they serve."*
>
> (Gabrielle Dolan, speaking to Gagel, 2019 Greatness Podcast)

I encourage you to continue learning and exploring the tools and techniques that allow you to be fully present with the people you lead and influence. As with many skills, being present starts with a desire to be present and a recognition that this may be an opportunity for improvement. Those around you will appreciate the effort.

> **Activity:**
>
> - Reflect upon your interactions with people for a week. Were you fully present? If not, what was causing you to be distracted? What could you do to be more present in the future?
> - Reflect upon other people and whether they are fully present to you. What could you say to help them be more present? Perhaps you could start with, "You seem a bit distracted, is there a better time to talk about this?"

Listening

Listening is hard. My friend Mary Rhinehart (who holds many significant positions, including former CEO and current Chair of Johns Manville) has had an incredibly successful career. She still sits on her hands at times during meetings to remind herself to listen. For Mary, this is a physical reminder to be patient and take the time to listen to the voices and opinions in the room.

> *"Stop talking for once, let other people speak. Women try to jump in with an idea and men often just keep talking."*
>
> (Salam-Haughton, 2023)

To listen is defined as "**1:** to pay attention to sound; **2:** to hear something with thoughtful attention: give consideration" ('Listen,' 2024). Thoughtful attention: two more great words to add to being focused and engaged. To be thoughtful means "**1a: absorbed in thought: MEDITATIVE; b: characterized by careful reasoned thinking; 2a: having thoughts: HEEDFUL; b: given to or chosen or made with heedful anticipation of the needs and wants of others.**" ('Thoughtful,' 2024). Another great phrase is, "made with heedful anticipation of the needs and wants of others" ('Thoughtful,' 2024). Sounds a bit like servant leadership. Listening entails not only physically hearing the words coming from someone's mouth but also focusing upon understanding the meaning and intent of these words. It involves "careful, reasoned thinking" while anticipating the needs of others ('Thoughtful,' 2024). This entails listening to understand not just the words but the thoughts and ideas that the other person is trying to convey, perhaps even thinking about the emotions that person is feeling. Are they excited about this idea? Nervous that it might not work?

To me listening begins with body language. Am I leaning forward as if interested, or leaning back as if bored? Am I looking into your eyes, or am I looking slightly to the side as if distracted? Am I writing down notes as you are speaking, a good indication that I am paying attention and interested in what you have to say? I can think of few meetings where I have not written something down. For me, to write is to process and to remember, and to respect the ideas being shared by others. Listening can sound like a simple act, but can you repeat back the idea that someone has shared with you? Do you find sometimes that someone is speaking, and you realize that you've missed a few things? I do.

> "*The ability to hear people, that's my gift, I know it's sort of my superpower. I hear when people say something to me, and if I don't think I have it, I'll say, 'Wait a minute, let me make sure I understand it.' I really hear people, especially with millennials versus Baby Boomers. I have a lot of millennials on my staff, and I think that's a generation that needs to be heard. It's a requirement for them and I really love being able to manage people in that sense of respect. That's my baseline for all management, I think it's just really hearing people and really allowing them to live in their circumstance. My lived experience is not yours.*"
>
> (Denise Burgess, speaking to Gagel, 2023 Greatness Podcast)

I believe great listening starts with really **wanting** to listen and the work you do to be present during conversation. Again, this is a skill that builds over time. Be kind to yourself when you think back on a conversation when you were not truly listening. There are courses you can take to improve your listening and research suggests this is an area for improvement for most leaders.

> *"There's a deficit in this space. Whilst the demand is high, the deficit exists very clearly from research upon research showing us that people are telling us that leaders aren't listening enough. They're not getting enough feedback. They're not getting enough one-on-ones to understand performance. They're just not connecting and engaging and showing the care, the compassion. And I think this is where humility comes into the connected leadership piece as well."*
>
> (Michelle Sales, speaking to Gagel, 2020 Greatness Podcast)

I personally relate to the topic of listening because it is a skill I continue to develop. The ability to listen is an incredible gift you give to those you lead and influence, and worth an investment of energy. I encourage you to continue to explore this topic.

Activity:

- On a scale of 1 to 10, rate your ability to listen. Why did you give yourself that score? What do you do well? What do you not do well?
- Think about your body language during an upcoming conversation. Lean forward, look that person in the eye, write down what they say. Then reflect – did this help me pay attention, be more focused, listen better?
- Consider asking for listening feedback after a conversation, "Hey, I'm working on being a better listener. Did you feel I was really listening to you during that conversation? Why or why not?"

Asking Great Questions

Friend and author Christy Belz once offered to complete a leadership circle evaluation for me (Leadership Circle, n.d.). As we were debriefing, I mentioned the fact that as a consultant I had never missed a client deadline. Her great question – "*At what cost?*" That question caused me to deeply reflect, to ask myself if I had met those client deadlines at the cost of my mental or physical health, at the cost of my relationships with my more junior associates because of how hard I drove them perhaps? It was the exact right question at the right time, and it caused me to think differently about myself.

> *"I think about the idea of going into a conversation where you just sit and ask questions and be curious. You've got to re-tap into that natural curiosity and*

> *find ways to change your environment up, or to change your disposition when entering a setting like a meeting. Challenge yourself to listen, ask questions, and resist the urge to offer your opinion and then be curious - was that difficult? Was it hard not to jump in and immediately answer and offer your own view? Or was it interesting to see who stepped up and what they shared when you came from a curious disposition?"*
>
> (Holly Ransom, speaking to Gagel, 2022 Greatness Podcast)

Great leaders ask great questions. One day as I spoke with Edgar about cutting back on one or more roles in my life, Edgar asked me, *"What would you miss most if you gave it up?"* A simple question but it reframed my thinking. That is what great questions cause you to do, to reframe your thinking.

Let's take a look at a few examples of being an "asking leader" versus a "telling leader" in our industry (Schein and Schein, 2021).

- A project engineer comes to you and says *"The framing crew that was supposed to come frame the rest of the 8th floor now can't be here for a week. What should we do?"* (this is also called throwing the monkey on your back) (Oncken and Wass, 1999).
  - You could "tell" them what to do, how to re-sequence the work to account for this challenge.
  - Or you could ask them great questions like:
    - o What are your thoughts on what we should do?
    - o Have you asked other people for their opinions? What do others think?
    - o What are the critical issues we should consider when coming up with solutions?
    - o What do you see as some of the possible solutions?
- A peer comes to you saying they would love to go for a new role in the company but are not sure they are qualified. What should they do?
  - You could go straight to offering your opinion to them – of course you are qualified, you should definitely go for it, etc.
  - Or you could ask them great questions like:
    - o What causes you to think you aren't qualified?
    - o Do you think the other candidates feel this way as well?
    - o What steps could you take to feel more confident about applying for the position?
    - o What will you regret more – going for it or passing up the opportunity?

Asking great questions helps people think. Great questions reframe and point out ideas that maybe we would not have come up with on our own. The trick is not to ask leading questions that instead of causing you to understand the other person are an attempt at swaying that person to your line of thinking. You must believe as a leader that this person has thinking to contribute to the issue at hand.

> *"Are you truly asking questions that you don't know the answer to, because so often we use questions as a way of making our argument. Humble inquiry is the opposite of that, it's that you don't have enough information to make an argument. Everybody has something they want to advise somebody else on. We often leap into advising people because we know some stuff and we think they maybe don't know some stuff, and so we are going to start using that currency,*
>
> *We're not asking you to change your personality from arrogant to humble. We're saying a whole different thing. Far be it from us to suggest to anybody that they change their personality, but humility in the here and now means, in any given situation, in a group, in a meeting, in a crisis, there's a lot you as an individual don't know, and there's a lot more that the group around you does know.* Humble Inquiry *is about accepting that 'here and now' humility which you could also call vulnerability if you're a Brene Brown fan. It means stepping outside of yourself and recognizing that the group around you collectively has more information and can make better decisions."*
>
> (Edgar Schein and Peter Schein, speaking to Gagel, 2021 Greatness Podcast)

One of the many personal profiles I have used throughout my career is the DiSC Profile, a low-cost, brief profile that helps me understand certain characteristics of myself and those I interact with (DiSC Profile, n.d.). My DiSC profile is high "D" and high "i." We "D"'s love to provide unsolicited opinions! Recognizing that trait in myself, I have developed certain go-to questions that invite more information and cause me to think and reflect versus jumping right into telling and giving my opinion. Friend, author, and advisor Mickey Connolly has helped me with several great suggestions as well (Connolly and Rianoshek 2002). Here are a few of my favorites:

- How does that make you feel?
- What caused you to think about?
- What are your viewpoints on this?
- How do you think others feel about this?

Technically this is not a question but more of a statement that Mickey taught me – *"Tell me more."* I find this easy to remember and applicable in almost any situation. It is an especially good phrase not only when you want to understand more, but also when you are trying to keep from having an immediate opinion or reaction to something. This phrase also helps me because I tend to ask one great question and then go off to the races in sharing my opinion or I move on to the next topic. I'm missing a significant opportunity to reflect upon what that person is thinking and to dive in more deeply as my mind races ahead.

> *"Another example of an interpersonal skill is the skill to balance asking and telling. Most of us make the mistake of doing a lot of telling, like, 'here's what I think and here's why.' We might forget to pause and say, 'But what do you think?' and 'What am I missing?' and 'How do you see it?'"*
>
> (Amy Edmondson, speaking to Gagel, 2019 Greatness Podcast)

The art of asking great questions is just that, an art. When interacting with people it is important to pause and reflect, to take a moment to think of that next great question. Our speed of operation in the construction industry does not allow for many pauses, but it should. To pause and reflect, to gather your thoughts and ask that great question. That is to be a great leader.

Activity:

- Think back to your conversations of the last week. Were you a "telling" leader or an "asking" leader?
- Is there as specific instance you can recall where perhaps an opportunity to ask a great question was missed? What was that great question?
- Think about the conversations and meetings you'll be in the balance of this week. What are some of the great questions you can think about to ask ahead of time?

Conclusion

Many years ago, I was facilitating a meeting of several large construction managers employed by a large tech company to discuss craft labor shortages, the impact on their projects, and related mitigation strategies. After our group dinner, Peter Nosler, co-founder of DPR, and I were sitting alone in chairs out by a deserted pool area. I can remember exactly where we were sitting and the deep, thoughtful discussion we had about construction and the world. I also remember thinking to myself, this person is leading a multi-billion-dollar construction company, and he has time to sit with me for 30–45 minutes to have a reflective, intelligent, thought-provoking conversation. What a gift.

> *"We are meant to be interested in each other because that's where creative ideas come from. That's where that spark of innovation might come from, when we're talking about something in an animated way and then all of a sudden, our conversation leads to an idea that neither one of us had when we walked in."*
>
> (Amy Edmondson, speaking to Gagel, 2019 Greatness Podcast)

As you reflect upon the information in this chapter, observe those around you. Are they present? Are they listening? Are they asking great questions? Even more importantly, what is the impact of what they are either doing or not doing on the people around them? Some of my greatest lessons in leadership have come from observing others. Habits are difficult to break. If one of these areas is a challenge, be kind to yourself. Don't try to focus upon everything at once. Consider small steps you can take to be more present and focused, listen with intention and learning, ask great questions. These skills are admired for good reason. These skills can be challenging. These skills make people feel important and heard, and that is an incredible gift you give to them.

References

Ackerman, C. E. (2018). 'How to live in the moment: 35+ tools to be more present,' *Positive Psychology,* 22 October. Available at: https://positivepsychology.com/present-moment/ [Accessed: 11 June 2024].

Connolly, M., and Rianoshek, R. (2002). *The communication catalyst.* Fort Lauderdale: Kaplan Publishing.

DiSC Profile. (n.d.). *What is DiSC?* Available at: https://www.discprofile.com/what-is-disc [Accessed: 11 June 2024].

Gagel, G. and Burgess, D. (2023). *Denise Burgess discusses her career as an African American woman leader in the construction industry* [Podcast]. 29 December. Available at: https://open.spotify.com/episode/77UCAi1FXUKeMOS6yfes8z [Accessed: 2 May 2024].

Gagel, G. and Dolan, G. (2019). *Gabrielle Dolan discusses her book Real Communication* [Podcast]. 21 May. Available at: https://open.spotify.com/episode/2By0s6SjK4tfZpXmXYB4GC [Accessed: 28 April 2024].

Gagel, G. and Edmondson, A. (2019). *Amy Edmondson discusses her book The Fearless Organization on psychological safety* [Podcast]. 30 October. Available at: https://open.spotify.com/episode/6joytIVhV3XlGFZx6rcodC [Accessed: 28 April 2024].

Gagel, G. and Goldsmith, D. (2023). *David Goldsmith, co-founder of the International Coaching Federation, discusses executive coaching* [Podcast]. 15 December. Available at: https://open.spotify.com/episode/1AktDJ16xXLuhPSr6ov1yG [Accessed: 2 May 2024].

Gagel, G. and Ransom, H. (2022). *Holly Ransom discusses her book The Leading Edge* [Podcast]. 30 September. Available at: https://open.spotify.com/episode/0M5G6yRwqvbBlKa302wmKb [Accessed: 1 May 2024].

Gagel, G. and Sales, M. (2020). *Michelle Sales discusses her book The Power of Real Confidence* [Podcast]. 11 December. Available at: https://open.spotify.com/episode/79XJIyGSMH9TVysaGJjA8V [Accessed: 28 April 2024].

Gagel, G., Schein, E., and Schein, P. (2021). *Edgar and Peter Schein discuss Humble Inquiry* [Podcast]. 26 November. Available at: https://open.spotify.com/episode/1xIf7ykZOWDXZMcrKvlId8 [Accessed: 1 May 2024].

Hansen, K. (2018). *Traction: The neuroscience of leadership and performance*. Sydney: EnHansen Performance.

Leadership Circle. (n.d.). *Leadership Circle certifications*. Available at: https://shop.leadershipcircle.com/ [Accessed: 6 July 2024].

'Listen.' (2024). *Merriam-Webster Online Dictionary*. Available at: https://www.merriam-webster.com/dictionary/listen [Accessed: 11 June 2024].

Oncken, W., Jr. and Wass, D. L. (1999). 'Management time: Who's got the monkey?' *Harvard Business Review*. Available at: https://hbr.org/1999/11/management-time-whos-got-the-monkey [Accessed: 11 June 2024].

Raypole, C. (2020). 'The beginner's guide to being present,' *Healthline,* 6 April. Available at: https://www.healthline.com/health/being-present [Accessed: 11 June 2020].

Salam-Haughton, S. A. (2023). Interview by Gretchen Gagel [Zoom], 26 October.

Schein, E. H. and Schein, P. A. (2021). *Humble inquiry: The gentle art of asking instead of telling*. Oakland: Berrett-Koehler Publishers.

'Thoughtful.' (2024). *Merriam-Webster Online Dictionary*. Available at: https://www.merriam-webster.com/dictionary/thoughtful [Accessed: 11 June 2024].

Thum, M. (2008). 'What is the present moment?' *Myrko Thum,* 31 August. Available at: http://www.myrkothum.com/what-is-the-present-moment/ [Accessed: 11 June 2024].

Ward, A. F., *et al.* (2017). 'Brain drain: The mere presence of one's own smartphone reduces available cognitive capacity,' *Journal of the Association for Consumer Research*, 2(2), 140–154. Available at: https://www.journals.uchicago.edu/doi/epdf/10.1086/691462 [Accessed: 11 June 2024]

Weinschenk, S. (2012). 'The true cost of multi-tasking,' *Psychology Today,* 18 September. Available at: https://www.psychologytoday.com/au/blog/brain-wise/201209/the-true-cost-of-multi-tasking [Accessed: 11 June 2024].

11

Utilizing Power to Influence

Introduction

When I moved to Australia in 2018, I realized that I knew little of Australian culture or the Australian construction industry, and I had no industry network. I thought about my personal brand and how I wanted to show up in this new country, and I thought about my "power" level. Did I want to arrive in Australia at a 10 out of 10 on power as the superwoman construction executive from America? Did I want to arrive in Australia at a 2 out of 10 on power, as in, "I don't know anything about how things work in the construction industry here?" Neither would have probably been good. I chose to come to Australia at about a 7 out of 10 power level, confident in the knowledge and wisdom I had gained during my 30+ year career in the US construction industry and humble about my lack of knowledge of the culture and construction industry in Australia. I believe my success in Australia is in part due to this thoughtful consideration of my power and how I exhibited it during my early time in Australia.

The word "power" often has negative connotations associated with it, probably because we relate it to abuses of power. But power is critical to our ability to lead and influence. We all have and wield power—some more than others, some for good, and some for bad. As a leader, it is critical to understand power and how it enables us to influence in a way that benefits ourselves and those we lead. I believe it is important as a leader to understand your natural level of power and how to regulate power not in a manipulative way, but in a way that reflects your understanding of the contextual environment of leadership. Effectively leading rests upon your ability to influence, and influence is not possible without power.

Building Women Leaders: A Blueprint for Women Thriving in Construction, First Edition. Gretchen Gagel.
© 2025 John Wiley & Sons, Inc. Published 2025 by John Wiley & Sons, Inc.

> As Dominique Gill, Founder and Managing Director of Urbancore, shares: *"Power can corrupt but it is also the force that drives change. I think as women in construction we are sometimes hesitant to think of ourselves as wielding power because of the negative connotations of the word. However, without power, progress stalls. If 'wielded' wisely, then we can drive the positive change our industry needs."*
>
> (Gill, 2024)

As you read this chapter, think about your natural power level and your ability to regulate your power and influence others. Reflect upon your sources of power. Are you cognizant of your power? How are you enabling the power of others?

Defining Power and Influence

Power is defined as "possession of control, authority, or influence over others" ('Power,' 2024). To influence is defined as "to affect or alter by indirect or intangible means" ('Influence,' 2024). As leaders, we are trying to influence others (team members, peers, superiors, etc.) to achieve some purpose or goal. It is via our relationships with others that we influence. We are essentially trying to affect or alter behavior—to convince someone to do things in a new way, or to do things to help the team achieve a goal. Our ability to influence is related to many elements of leadership (our ability to build trusting relationships, to listen, etc.). Understanding power is a key element of that.

> Often as leaders, we influence people without a title or direct authority. I am reminded of working with Procter & Gamble to implement Lean construction practices globally, and the leadership and influence required by retired Associate Director of Global Capital Management for Procter & Gamble Mike Staun to make this happen.
>
> *"I think some people thought I was hijacking our capital management leadership team even though I was the leader of it, a small group that reported directly to me. But most of the people that were going to be needed to really make this transformation, to make this move, were people who were in their businesses. They didn't report directly to me. For the most part I had to be in an influence mode. It was really a matter of working with those people and saying, 'Hey, there's a better way to do this', and it took a while. The key was continuing to go back to 'this is how we want to work' versus 'this is how we're currently working', and then getting people to say, 'okay, we're willing to take a chance with this, let's go ahead and do a pilot project here'; and then you get a pilot, and another pilot project, and another pilot project. Finally, you build up a little bit of momentum."*
>
> (Mike Staun, speaking to Gagel, 2019 Greatness Podcast)

Think of the leaders you admire in your organization and in the construction industry. Perhaps you admire them because they know a great deal about how to successfully execute construction projects. Perhaps you admire them because they are good at building relationships. Perhaps you admire them because of the positions they have achieved. Your admiration probably means these people are viewed by you as leaders who influence your thinking, and that in part comes from their sources of power.

French and Raven's 1959 research identified six sources of power at work that I believe to be appropriate even in today's modern business context (Kovach, 2020).

- **Legitimate Power** – This is the power you legitimately hold because of your position or title. When I was the CEO/President of WFCO, the Chair of the Board asked me to complete a leadership assessment process. One important takeaway for me was that positional power always exists for the person with the title, and it should be used sparingly as a leader. I always had the power of being the CEO/President of the organization. To rely heavily upon positional power is analogous to constantly whipping the racehorse. Great leaders speak last and typically use positional power as a last resort.

  That is not to say that using positional power is not appropriate at times. While leading one company, our partners decided we were going to implement a new travel expense app to streamline the process of capturing client-related travel expenses. After six months, one partner was still not using the app and it was wreaking havoc on our invoicing and administrative processes. After asking nicely several times, I used my positional power and said, "If you do not use the app, your travel expenses won't be reimbursed." Using my positional power as CEO was my last (and appropriate) approach.

- **Reward Power** – This is the power you have to reward people for their efforts to achieve the desired results of the team and organization through tactics like praise, compensation, and bonuses. As a university professor, I have reward power, as I can reward students with good grades for their efforts. As a leader, it is often more effective to use a carrot than a stick, as rewards can create greater influence than punishments. It is important to understand that different people are motivated by different rewards. One person may value praise as a reward, whereas another might be motivated more so by monetary rewards.

  My flight was delayed while traveling to my first leadership meeting for one organization, and I walked in a few minutes after the meeting had started. It was only after the first break that I realized the President handed each member of the leadership team a $100 bill after every break. Why? Because it was a difficult group to keep focused, and he wanted everyone in their chair and ready to contribute on time. It worked! When I was leading the Coca-Cola bottling facility in Denver, Colorado, I would beg the sales manager for Denver Broncos or Rockies tickets that I would then give team members as a reward for productivity. Not only did they receive a "prize," but they also knew I cared enough about them to secure the tickets.

- **Reverent Power** – This is the power we gain by creating effective relationships whereby people admire us as leaders, peers, and teammates. People may appreciate us because of our purpose, our values, our outcomes, and our results. They admire us for these attributes, and this creates the power to influence their behavior.

  I am reminded of my relationship with my friend and mentor Barb Grogan. My admiration for Barb as a leader in the community—a caring, involved citizen of the world—means she has the power to shape my behavior. My coach and friend Edgar Schein is another person who had reverent power in our relationship. I deeply admired Edgar not just for his research and consulting experience, but also for the person he showed up as every time we met.

- **Expert Power** – The knowledge and skills you draw upon to execute your role make up your expert power. People may view you as an expert on a variety of subject matter topics (such as engineering, project management, or Lean practices), and even leadership topics (such as how to help teammates resolve conflicts, how to motivate one another, or what to do when a colleague is not performing). Your knowledge is a significant source of power.

  An excellent example of this is Jennifer Briggs, one of my fellow Board members at Brinkman Construction. We brought Jennifer onto the Board because of her significant experience with ESOPs—an area where I have much less experience. When we discussed an ESOP-related strategy, I deferred to Jennifer and her knowledge. Jennifer had "expert power" to influence my thinking. Teresa Magnus also shared with me during her interview for the book that her law degree provides her with expert power—especially among men.

- **Informational Power** – This power occurs when people have access to information that other people want and need. I think of some of the long-term employees I have met who not only have explicit information but also the implicit knowledge they've gained over the past 35 years—information that isn't written down and yet is invaluable to people's success.

  Back in the 1990s, I started conducting a construction industry study, first with the Construction Management Association of America and later with the Construction User Roundtable. This study involved interviewing 40 or so of the top client leaders in the world from such companies as Procter & Gamble, General Motors, and Microsoft to determine the key trends in the industry, challenges we are likely to face in the coming decade, and solutions that people were utilizing to mitigate these challenges. At the time, I was starting the Owner Services Group at FMI, and our team did not have informational power in this market. The study provided my team with both brand recognition and informational power because we had gathered information that was invaluable to others.

- **Coercive Power** – This power centers around the use of threats to achieve desired behaviors and outcomes. Although many of us as leaders are hesitant to admit that we use this, we do, and coercive power is appropriate in some instances.

  While President of an organization that had two distinct brands, I was frustrated that our teams were often using just the first common word of the two brands in proposals, without adding the differentiating words of the brands. I mentioned this to our team a few times, but the problem remained. Finally, during one of our all-team meetings, I said I was going to start fining the author $100 every time I saw the first brand word without something after it. It worked! This is not a power that a leader should use frequently, but it is appropriate at times.

Deliberating considering and utilizing our sources of power is an effective way to think about your ability to influence others as a leader. Without power of some type, there is little influence. I encourage you to be aware of your sources of power and how you wield this power to influence and lead.

Activity:

- Make a list of your current sources of power. How effective are you at appropriately using each of these sources of power? Where might you improve? What do you believe others would say about your level of power?
- What other sources of power might you work to develop? What steps might you take to do so?
- Before an important conversation or meeting, mindfully consider the level of power you want to bring into this meeting and why. Afterward, reflect on your effort. Do you think you achieved the desired level of power? Why or why not? What could you have done differently?

Regulating Our Power Level

When I left the construction industry to become the CEO/President of WFOC, I entered into the extremely different context of a nonprofit setting with an entirely female staff. I was accustomed to the "9 out of 10" power level necessitated by the culture at FMI—male dominated and often a struggle for resources and respect. I considered my sources of power in this new role within a new industry. I had much less expert power, as I had never led a nonprofit organization. I had positional power, but I knew to use that sparingly. I consciously brought my power level down to about a 6–7: not too powerful, but also powerful enough to command the respect of my team. I worked hard to build the relationships necessary to have reverent power, and I implemented a bonus compensation plan that also provided me with reward power. As I learned more about public policy and our work, I gained expert power.

I can think of two instances in the first six months of that job where I did not assume the correct power level. In the first instance, I became "triggered" by a volunteer (the chair of our Fundraising Committee) who made several comments during a meeting about my lack of fundraising experience. I do not remember exactly what I said, although it was probably something underscoring my past experience in selling large consulting deals and this being "not my first rodeo." I clearly remember this person's reaction—not good. I later apologized and we became close friends, but my efforts to control my level of power had let me down. I had used too much positional power while lacking reverent power in this relationship.

I also had the opposite experience, where I walked into a meeting with my power level too low. I was meeting with one of my new peers in the foundation community to discuss collaboration amongst foundations in Denver, as I had no experience with this. During our meeting, I mentioned that I was grateful for her help and that I had much to learn. After saying this again near the end of our meeting, this person said to me, "*If you have so much to learn, why did they hire you?!*" I lifted my power and confidence a bit and articulated the many skills I brought to the position, increasing my expert power by stating how these skills were relevant in a new context. This was another important learning experience for me—that I had entered into this meeting with too low a power level.

In the construction industry, understanding the effective use of power and being able to regulate your power are critical. I remember the first time I walked into a Brinkman Construction jobsite trailer for a project walk-through as Chair of the Board. I chose to have my power at about an "8," confident but not overly so, focused on expert power and hoping to build reverent power. I asked thoughtful, relevant questions during the team's presentation that demonstrated my knowledge of construction and my respect for their work as a team. I will always remember the end of that meeting. As we paused before heading out to walk the project, the Project Superintendent turned to me and asked, "*Where did we find you? Or how did you find us?!*" To me, it was a sign of his respect for me, and his unconscious bias that few women with my construction experience exist in the world.

Consciously thinking about how we show up in a certain context is not disingenuous. Thoughtfully preparing for each unique situation as a leader demonstrates respect for that specific situation and which behaviors are most appropriate to achieve your desired outcomes. I have tremendous respect for a leader who can step into their positional power in a crisis or utilize their knowledge power for the betterment of the team. I try to be conscious of my desired power level in each situation I enter into. Do I want to be a "6?" An "8?" What sources of power am I relying upon? How am I trying to influence this person? Being attuned to your power and influence is an important element of great leadership.

Activity:
<ul><li>What is your natural level of power on a scale of 1 to 10, with "1" being low and "10" being high? How effective are you at moderating this level of power? How might you practice moderating your level of power over the coming weeks?</li><li>Think of an instance where you might have benefited from more or less power. What could you have done differently to moderate your level of power?</li><li>How will you be mindful of your power in the future?</li></ul>

Conclusion

I stood on a sidewalk in Melbourne in thoughtful contemplation. I was about to meet with someone face-to-face for the first time; someone I respected, would be working with, and hoped would respect me. I was joining the team of a global engineering firm as an outside advisor and was expected to contribute to their thinking on the global construction industry. I thought about my sources of power—no positional power, and no reverent power (new relationship), but I did have informational power about a tangential market for this organization. I was being brought on as an independent expert; what was the appropriate power level? Too little, and he would question his choice. Too much, and I would be viewed as a brash know-it-all. I decided on "7," and it seemed to work.

Consider the sources of power you currently have or aspire to—your natural level of power and your current ability to regulate that power. Be aware of your power and how you come across to people in a way that is authentic to you AND commands the appropriate respect for you and your ability to influence others as a leader. Power is a necessary ingredient of leadership because without power, there is no influence, and without influence, no one is following.

As Ernest A. Drott, Chief, Engineering and Construction Division, US Army Corps of Engineers, Great Lakes and Ohio River Division, shares: "*I have two daughters and I want women to have opportunities. Women must be tough to survive in our industry, must often do better than their male counterparts* to move up. I encourage women in our industry to know their craft, look for leadership opportunities, and focus on their ability to influence others."

(Drott, 2023)

References

Drott, E. A. (2023). Interview by Gretchen Gagel [Zoom]. 27 October.

Gagel, G. and Staun, M. (2019). *Mike Staun discusses his work at P&G to lead transformational change* [Podcast]. 9 December. Available at: https://open.spotify .com/episode/3eglXD5HdZMRDGmo2JJEce [Accessed: 28 April 2024].

Gill, D. (2024). Interview by Gretchen Gagel. 10 July.

'Influence.' (2024). *Merriam-Webster Online Dictionary*. Available at: https://www .merriam-webster.com/dictionary/influence [Accessed: 11 June 2024].

Kovach, M. (2020). 'Leader influence A research review of French & Raven's (1959) *Power dynamics,*' *The Journal of Value-Based Leadership*, 13(2), 15. Available at: https://scholar.valpo.edu/jvbl/vol13/iss2/15/ [Accessed: 11 June 2024]

'Power.' (2024). *Merriam-Webster Online Dictionary*. Available at: https://www .merriam-webster.com/dictionary/power [Accessed: 11 June 2024].

12

Leveraging Emotional Intelligence

Introduction

My first year as CEO/President of an organization, we set aside a training budget for everyone to focus on an area of self-development. Unfortunately, many people were too busy to devote time to this, and the budget went unspent. The next year, I decided to spend the entire budget for everyone, and I hired my friend Brent Darnell to coach our team on emotional intelligence. Brent has devoted much of his life to helping people in the construction industry become emotionally intelligent and more inclusive, and he understands the unique aspects of our industry (Darnell, 2019). It turned out to be one of my better decisions as a leader. He has now been my emotional intelligence coach on-and-off for 10 years, and I just opened my copy of *The People Profit Connection* to Brent's inscription: *"To Gretchen, an enlightened woman with high EI. Glad to know you. Brent"* (Darnell, 2019). Thank you, Brent, for the vote of confidence!

I have included emotional intelligence in this portion of the book because while I agree that the development of emotional intelligence is an inwardly focused "ME" activity, the result of high levels of emotional intelligence is strong relationships. Emotional intelligence is a critical leadership skill for building relationships because one emotionally triggered reaction can completely erase the good will and trust you have developed with a person in minutes, and one moment of ignoring the emotions of another can damage that relationship irreparably. I know because this has happened to me, and I have deeply regretted it each time.

> *"When you're the CEO of a company, you do have a ripple effect. The coefficient of your impact is far greater by the nature of the role you have, so you have a different responsibility in terms of what are you telegraphing, how are you handling that set of emotions, whether constructive or not constructive? You're angry, but rather*

Building Women Leaders: A Blueprint for Women Thriving in Construction, First Edition. Gretchen Gagel.

> *than raging, do you need to take a private moment, do you need to reschedule that meeting? What's the more constructive path forward?"*
>
> (Amy Jen Su, speaking to Gagel, 2021 Greatness Podcast)

The good news is that emotional intelligence can be learned, and we are not "hard-wired" to a certain level of emotional intelligence like we are with our IQ or other skills, such as being a detail-oriented person. I have taken the emotional intelligence test three times and seen my progress. Am I perfect? No, but I am better. We also need to realize that showing your emotions is not a sign of weakness. *"Some people confuse leadership with being invincible and knowing all the answers. You are also human. You must show your emotions so they know they can be human. It's okay to feel fragile and frustrated. I use these emotions to make me better"* (Gomez Fabra, 2024).

As you read through this chapter, I encourage you to reflect upon yourself and others, how you and others have displayed strong emotional intelligence, and perhaps scenarios where more emotional intelligence would have been helpful. Think about your level of awareness of your emotions. Do you control your emotions? Or do they control you?

Defining Emotional Intelligence

Daniel Goleman is regarded by many as the "godfather" of emotional intelligence stemming from the emotional intelligence framework shared in his 1995 book, *Emotional Intelligence: Why It Can Matter More Than IQ* (Goleman, 2005). While many others have expanded upon the thinking on emotional intelligence, I focus on three important aspects of emotional intelligence from Goleman's model in this book: (1) self-awareness – our ability to feel and understand our emotions; (2) self-regulation – our ability to regulate our emotions; and (3) social awareness – our ability to sense and have empathy for others' emotions (Goleman, 2005). I believe focusing on these three elements creates a foundation of the emotional intelligence that is critical to strong relationships and great leadership.

Emotional Self-Awareness

Nearly every day that I am in Melbourne, Australia, I jog in the beautiful Fawkner Park where many young children play. I will often hear a parent console a child who has fallen by saying, "Don't cry." One day, I experienced what appeared to be a grandmother consoling a small boy in her lap by repeatedly saying to him, "Be happy," while he was screaming, "But I don't want to be happy!"

Many of us are conditioned during our upbringing to suppress our emotions—to push away sad feelings and show up as "okay," even when we are not "okay." When I ask leaders what we are taught about emotions at work, I hear that we are encouraged to always be upbeat and are not encouraged to share our emotions—especially if these emotions are sensitive. When I ask leaders what we **should** be saying about emotions at work, I hear things like, "Appropriate emotions are okay," "It's okay to show my emotions," and "It's okay to not be happy every day." Of course, there is an appropriate way to express our emotions. I believe that pretending everyone is 100% okay every day and that emotions are to be suppressed can cause a great deal of stress and anxiety.

I also believe this topic is even more challenging for women in the construction industry because we can be negatively labeled as "emotional" when the ability to be aware of and appropriately express our emotions is a good thing. Fortunately, I believe this is changing, and it takes people with tremendous courage to help us understand that it's okay to feel emotions. I am a huge fan of tennis great Serena Williams who, on November 29, 2023, posted on Twitter, "I am not ok today. And that's ok to not be ok. No one is ok every single day. If you are not ok today I'm with you. There's always tomorrow. Love you" (Williams, 2023). I think this message is important, given the perceptions perpetuated by social media that "everyone is okay every day and doing better than me."

> "*I think we've come a long way from when we were growing up, when we were told it isn't business, it isn't personal, keep your personal life separate, it's not emotional, don't be so emotional and all this. I think that's old school leadership and there's a growing recognition that business is personal because we're all humans and it's okay to show emotion. If you want people to be passionate and you want them to be themselves, it's okay to show emotion.*"
>
> (Gabrielle Dolan, speaking to Gagel, 2019 Greatness Podcast)

We also tend to think about our emotions at a very high level—I am happy, or I am sad—versus digging more deeply into the nuances of our emotions. Plutchik's Wheel of Emotion is an effective tool for helping us explore our emotions in more detail (Figure 12.1) (Karimova, 2017). It is helpful to ask ourselves, "What am I truly feeling in this situation? Disappointed? Relieved? Worried? Frustrated? And why am I feeling this way?" One day, I walked into a meeting feeling really fired up about a project, and I left feeling "blah." I thought to myself, "Am I feeling challenged? Frustrated?" By exploring my emotions, I realized I felt "dismissed" and why a particular person had caused me to feel that way. I was able to sit down with this person a couple of days later and provide them with specific feedback that they appreciated because I was able to pin down the emotion I was feeling and what actions had caused me to feel this way.

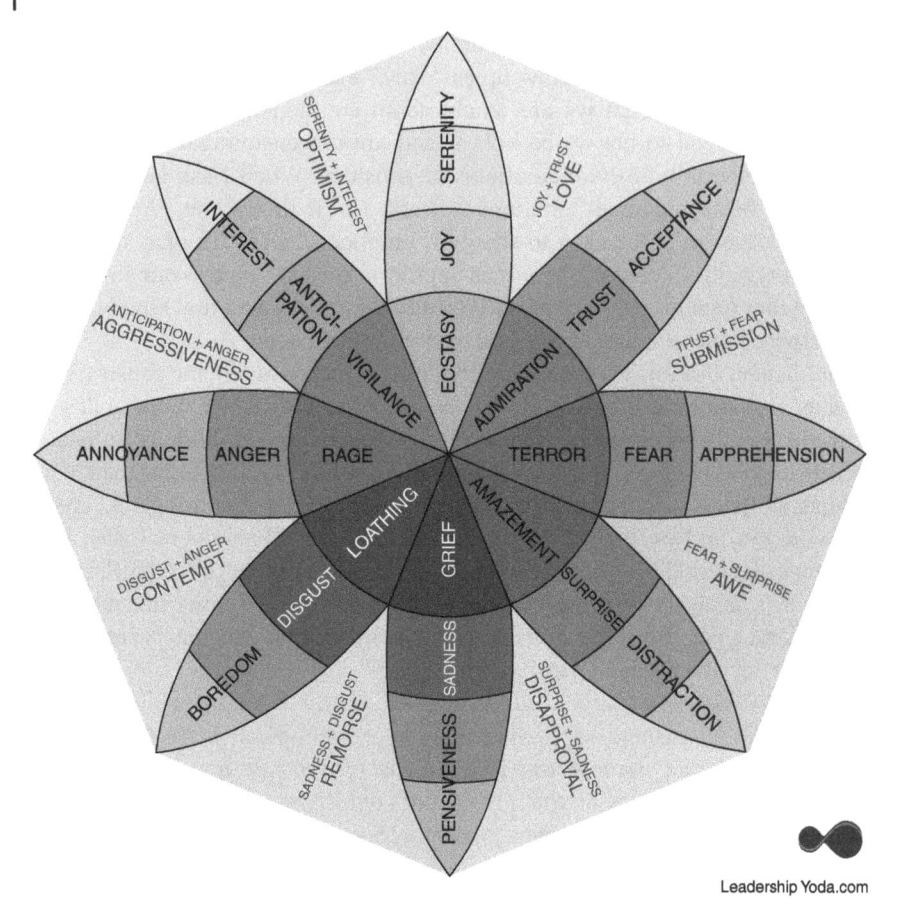

Leadership Yoda.com

Figure 12.1   Plutchik's Wheel of Emotion

Emotional intelligence begins with self-awareness—an ability to feel our emotions, to understand and describe our emotions in detail, and to understand why we are having these emotions (Goleman, 2005). This, in turn, leads to the ability to be in control of our emotions instead of having our emotions control us.

Activity:

- For the next week, stop to reflect upon the emotions you are feeling.
- Try to describe that emotion in more detail using Plutchik's Wheel of Emotion. Be present to the emotion as you feel it.
- Dig more deeply into why you are feeling that emotion. What actions or behaviors by others are causing you to feel these emotions? To what level are you feeling these emotions?

Emotional Self-Regulation

Several years ago, I was interviewing people who worked for a construction leader, and I started to sense that this person was not highly effective at regulating their emotions. One of his direct reports said to me, "So-and-so is great most of the time, but then there are moments of 'rage mode.'" Rage mode?! Not a good thing when you are working to build strong, trusting relationships. The ability to not only understand our emotions, but also to control our emotions and knee-jerk reactions that damage relationships, is a critical leadership skill. Daniel Goleman refers to this as "emotional self-regulation" (Goleman, 2005).

The ability to regulate our emotions begins with the ability to feel where our emotions bubble up in our bodies, especially when we are being "triggered"—the term used to describe an "amygdala hijack" (Goleman, 2005). The amygdala is an almond-shaped sliver of our brain that is part of the limbic system and plays a critical role in how we feel emotions (Hansen, 2018). You have probably heard the saying "Fight, Flight, or Freeze." One of the main functions of our brain is to keep us alive, and the amygdala is one of the stars of this show, helping us understand when we should be afraid. I have learned that social and physical fear feel much the same in our brains, so if we are fearful of missing a deadline or of our inability to learn a new technology, the amygdala goes to work (or triggers), starting what many call an "amygdala hijack" (Goleman, 2005).

> "The amygdala is a part of the brain that processes emotions as it responds to fear and organizes the fight-flight-freeze response when there is a threat in our environment. That threat could be a nasty email or a forgotten deadline or it can be a car pulling out in front of us. Sometimes we take a bit of a low-road communication or reaction to an event. We might speak harshly, or we may freeze and not be able to speak at all. Sometimes it's not our finest moment and we may regret that we either said or behaved in a particular way.
>
> "The amygdala organizes blood flow away from the prefrontal cortex to be in that fight, flight, or freeze mode. It's headed south to the heart, lung, and limbs to be able to react. The blood has moved away from the prefrontal cortex which is our thinking part of our brain where we do our executive functions, our higher order of thinking, our planning, and strategizing, and understanding, and recalling, and even behavior modification. When the blood is not there, we are not able to do those functions as well."
>
> (Kristen Hansen, speaking to Gagel, 2019 Greatness Podcast)

Where do I feel my emotions when my amygdala is being triggered? In my shoulders. When I feel my shoulders near my ears, I know I need to step back and take a deep breath. Going back to the story of being in a construction trailer with 10 men and my work colleague forgetting to introduce me, at the time, I could feel my shoulders rising as I became more and more frustrated. What was I feeling? Anger? Probably disappointment that I had tried to avoid the situation by discussing our preparation the prior night, failure at not conveying my needs adequately, embarrassment that I was sitting there, and that these people had no idea who I was or why I was there.

Beyond understanding where I feel triggered in my body, I have also found it helpful to reflect upon and understand **what** triggers me. It took me years to figure out that because I love to have plans, someone disrupting my plan really sets me off. I juggle many balls in my life (probably too many at times), and my plan is the way I stay organized and reduce my stress. I also become triggered when someone challenges either my professional integrity or my ability to do something well. You might think, "Well, those are things we are all triggered by," but I am not talking about just getting a little upset; I am talking about becoming really triggered and potentially having an inappropriate emotional response. I encourage you to think about what triggers you, as that will help you preempt triggered behaviors which may damage relationships.

It is also important to understand that when we are under sustained stress, our amygdala is more likely to trigger. "*If the amygdala is triggered regularly through ongoing and chronic stress or through a traumatic event then the amygdala responds by being more hypervigilant. We tend to see this in some managers that are probably burnt out or have been chronically stressed for some time and certainly there are a lot of mental health issues associated with an enlarged amygdala. Post-traumatic Stress Disorder (PTSD) is essentially an enlarged amygdala*" (Gagel and Hansen, 2019/ with permission of Kristen Hansen). I experienced this personally in 2018 when I was winding up my role as CEO of Continuum Advisory Group and defending my PhD dissertation 20 days before moving to a new country. I now realize that I was not living the best version of myself as a leader during this period. Unfortunately, the construction industry is a hotbed of sustained stress, and this has a tremendous impact on our people.

> "*They're under a lot of stress and they don't get a lot of sleep, and they eat crappy food, and they don't exercise, and then we expect them to perform at a high level. Well, that's just insane. It's a testament to the people in the industry they do perform at a high level, and they finish the projects. But guess what? It comes at a really high cost because you see people in the industry, they're overweight, they're having health issues, they're having heart attacks, they have diabetes, and we have the highest suicide rate of any industry. Look at some of the guys in the trades, they look older than they are. Lots of times they're addicted to nicotine and caffeine and alcohol. It's a part of the culture of this industry.*"
>
> (Brent Darnell, speaking to Gagel, 2020 Greatness Podcast)

The opposite of a hypervigilant amygdala exists as well. In the documentary *Free Solo*, climber Alex Honnold successfully climbs a 3,000-ft vertical cliff, El Capitan, without ropes (*Free Solo*, 2018). During the documentary, Alex is given an fMRI which reveals that his amygdala is calmer than a normal person's, and Alex credits having developed this through years of meditation (Sharma, 2022). We can train our amygdala to be calm in the same way we train our muscles. Emotional self-regulation allows you to catch yourself when you are being triggered, to be present to this happening, and to hit the proverbial "pause button" to stop the negative reaction before the damage is done (Goleman, 2005). To be adept at this, you must first be able to feel the emotion, know the emotion is happening, and feel it in your body.

> "*One of the things that I think is useful to recognize is that our response to threats, any form of threat, is a combination of our amygdala alerting us to that threat but also our ability to regulate the amygdala through our prefrontal cortex. Our prefrontal cortex houses our ability to inhibit our response or control our emotions, so it's a very important part of that two-way relationship. If we don't have strong regulatory ability, then our amygdala can go off half-cocked. But we can train ourselves to have greater ability to regulate that emotion.*"
>
> (Kristen Hansen, speaking to Gagel, 2019 Greatness Podcast)

Techniques such as mediation can help us train our brain to regulate our emotions. Other ways to create a "pause button" between feeling the emotion and acting include:

- Taking deep breaths.
- Leaving the space for a few minutes.
- Asking a "pausing" question.
- Talking to someone else (perhaps even venting a bit).
- Journaling about your emotions.

For me, deep breathing works best. I have also realized that it is okay to say, "I'm having a bit of an emotional reaction to this conversation, can we come back and talk about this again tomorrow?" or "I need to think about this—can we revisit this later?"

The ability to pause **before** doing something or saying something that will irrevocably damage your relationship with someone you lead or influence is a critical skill in effective leadership. This is not about not having emotions. On the contrary, feeling our emotions, exploring our emotions, being present to our emotions, and sharing our emotions are a good thing. What is not good is having a knee-jerk, emotional reaction to something that damages a relationship. You can still respond with emotion, but we need to ensure that we are controlling the emotions instead of the emotions controlling us.

Activity:

- Think about times that you've been "triggered." Where did you first begin to feel this in your body?
- When you have been triggered, what were the primary causes?
- Reflect upon the list of activities you have used to regulate your emotions in the past. What has worked? What hasn't worked?
- Given this discussion, what activities might you try in the future?

Emotional Social Awareness

I had a difficult conversation with a person I collaborated with on a significant project as I tried as kindly as possible to explain that our differences in style were causing me a great deal of stress. At the beginning of the conversation, I was focused intently on my needs and emotions—"I need to feel less stress, I am feeling frustrated, I need this person to understand why I do not want to collaborate with him in the future." Then I began to realize that this person was having an emotional reaction, and I started to focus on their emotions. I thought to myself, "Their feelings are hurt," and I started to emphasize, to relate this to similar experiences I have had. Perhaps this person was feeling less adequate, less confident, and maybe even angry and frustrated? By considering their emotions, I was able to alter my comments in a way that achieved my needs, honored their feelings, and salvaged our personal relationship.

Social awareness is the ability to understand and empathize with the emotions others are feeling (Goleman, 2005). Social awareness is deeply rooted in our ability to be empathetic to others. I find empathy to be another of those complicated words that is used frequently but perhaps not deeply understood. Empathy is defined as "the action of understanding, being aware of, being sensitive to, and vicariously experiencing the feelings, thoughts, and experience of another of either the past or present without having the feelings, thoughts, and experience fully communicated in an objectively explicit manner" ('Empathy,' 2024). When was the last time you or I sat in a project update meeting and tried to thoughtfully experience the feelings of others at the table? Did we look around the room and think, *"This person is feeling anxious about this report-out,"* or *"This person is feeling disappointed in how this project is going?"* As industry leaders, we tend to focus on metrics, outcomes, tasks, and tactics—anything but emotions.

Why is empathy so important? For most people, knowing you care as a formal or informal leader is critical to the depth and quality of your relationship, and better relationships mean better individual and team outcomes. The ability to sit with someone with their emotions, not to necessarily try and fix things but to be deeply empathetic to what someone is feeling, is an incredible gift. People want to be understood and cared about. People do not follow leaders who do not care about them, and to not demonstrate empathy is to be seen as a leader who does not care.

Activity:

- Reflect upon your recent interactions at work. Were you present to the emotions others were experiencing? How could you have focused more on understanding their emotions?
- Think about yourself and your ability to demonstrate empathy. How might you improve upon this in the future?

Conclusion

One day, as I was boarding a flight from Melbourne to Sydney, the scanner made that weird beeping noise that signals something is wrong, and the gate agent said to me, *"I'm sorry, you're not on this flight."* To provide even greater context, this was the second flight I was taking after the end of the COVID lockdowns, and my amygdala was probably the size of a small orange! I looked at my boarding pass and asked her if I had the right gate for the right flight number, to which she responded, *"Yes."* I showed her my boarding pass for that flight. She again said, *"You're not on this flight,"* and I felt my shoulders rising. You see, I had a plan—fly to Sydney, dinner with a client, a good night's sleep, several meetings the next day. This flight was critical to my plan, and I know that when someone messes with my plan, it triggers me! Especially when she said, *"Please step aside, you'll need to take a later flight while we sort this out."* No, that is not the plan!

I felt my emotional reaction in my shoulders, and caught myself. I took three deep breaths, smiled, and calmly said, *"It would appear that I have a boarding pass for this flight, and something has happened. It would be really helpful if we could sort this out in time for me to catch this flight."* The gate agent responded well and asked me to step aside while she radioed her supervisor. For some reason, I had been moved to a flight to Canberra (not my plan), but she was able to put me back onto the flight just as the door was closing. Mission accomplished! Had I had an emotional reaction, the gate agent probably would not have worked so quickly to resolve the issue. This is a simple example of my emotional intelligence at work.

I started this chapter by saying that emotional intelligence can be learned, and it can. My suggestions to you:

- Read and reflect upon emotional intelligence.
- Take an online course in emotional intelligence.
- Hire an emotional intelligence coach.
- Choose a skill to work on, and practice it.

Emotional intelligence is critical to your success as a leader—a core skill that you can build and is worth focusing on.

References

Darnell, G. B. (2019). *The people profit connection: How to transform the future of construction by focusing on people.* 4th edn. Atlanta: BDI Publishers.

'Empathy.' (2024). *Merriam-Webster Online Dictionary.* Available at: https://www.merriam-webster.com/dictionary/empathy [Accessed: 11 June 2024].

Free Solo. (2018). *Directed by E. C. Vasarhelyi.* Available at: National Geographic [Accessed: 11 June 2024].

Gagel, G. and Darnell, B. (2020). *Brent Darnell discusses his book The People Profit Connection on how to take great care of people* [Podcast]. 6 January. Available at: https://open.spotify.com/episode/37vfyvki0GqeeoVJn02CdD [Accessed: 28 April 2024].

Gagel, G. and Dolan, G. (2019). *Gabrielle Dolan discusses her book Real Communication* [Podcast]. 21 May. Available at: https://open.spotify.com/episode/2By0s6SjK4tfZpXmXYB4GC [Accessed: 28 April 2024].

Gagel, G. and Hansen, K. (2019). *Kristen Hansen discusses her book Traction: The Neuroscience of Leadership* [Podcast]. 29 July. Available at: https://open.spotify.com/episode/1ayV1cm3IedLOpHKA787Ka [Accessed: 28 April 2024].

Gagel, G. and Su, A. J. (2021). *Amy Jen Su discusses her book The Leader You Want to Be* [Podcast]. 12 November. Available at: https://open.spotify.com/episode/4ePtBXmgN40IG7S7pKiFu6 [Accessed: 1 May 2024].

Goleman, D. (2005). *Emotional intelligence: Why it can matter more than IQ.* London: Bantam.

Gomez Fabra, M. P. (2024). Interview by Gretchen Gagel [Zoom]. 24 January.

Hansen, K. (2018). *Traction: The neuroscience of leadership and performance.* Sydney: EnHansen Performance.

Karimova, H. (2017). 'The emotion wheel: What it is and how to use it,' *Positive Psychology,* 24 December. Available at: https://positivepsychology.com/emotion-wheel/ [Accessed: 11 June 2024].

Sharma, S. (2022). '"My brain is different": American rock climbing legend Alex Honnold once found a unique way to mitigate risks in free soloing,' *Essentially Sports,* 27 December. Available at: https://www.essentiallysports.com/us-sports-news-olympics-news-rock-climbing-news-my-brain-is-different-american-rock-climbing-legend-alex-honnold-once-found-a-unique-way-to-mitigate-risks-in-free-soloing/ [Accessed: 11 June 2024].

Williams, S. (2023). [Twitter]. 29 November. Available at: https://x.com/serenawilliams/status/1729557645437046976 [Accessed: 11 June 2024].

13

Building Trust and Psychological Safety

Introduction

Back in the 1990s, I had the good fortune to work on the construction of the NFL Baltimore Ravens team stadium with the remarkable Alice Hoffman, Project Executive for the client. Alice approached me with her vision of creating a highly collaborative, trusting project team capable of achieving outstanding results, and I was excited to help her achieve this vision. Our first step was to take the senior executives from the client, the construction management team, the architect, the key subcontractors, and the major engineering firms out into the woods for a day of Outward Bound, a highly experiential outdoor learning methodology that brings together people to explore leadership and teams (Outward Bound, n.d.). One of the exercises that day was a "trust fall," where a member of the team climbs onto a platform and falls backwards several feet into the extended arms of the other members of the group, hopefully unharmed! The trust that was developed that day, and over two subsequent days of teambuilding meetings, created a strong foundation that continued throughout the construction project. It was an amazing experience to see Alice's vision of trust and collaboration come to life and experience a construction project team that achieved outstanding results.

I have paired psychological safety with trust in this chapter because I believe the two go hand-in-glove. A leader who creates psychological safety has given people the gift of being able to take risks without fear of failure, point out issues without fear of repercussions, and throw out crazy ideas without fear of ridicule (Edmondson, 2018). This mitigation of fear is builds and is built by high levels of trust.

> *"We live in a world in which learning and innovating and teaming are more important than ever. Innovation is crucial to the success of most companies. Teamwork is also crucial to the success of most companies. Organizations don't last long when people aren't actively learning. All three of those things - learning, teaming, innovating - require psychological safety."*
>
> (Amy Edmondson, speaking to Gagel, 2019 Greatness Podcast)

As you read this chapter, think about your skills in building trust and how they might be heightened. Consider the level of psychological safety on your team and how you might improve it. You may also consider how you continue learning about these two important topics.

Defining Trust

I once worked with a construction project where there was an extremely low level of trust within the team. Through speaking with the various project team members, I gleaned that several of them had experienced low-trust relationships on prior projects that led to poor performance, and they had an attitude of, "I'm not letting that happen to me again." As we worked through this challenge and built strong trust within the team, I learned a great deal about the fragility of trust and how our past experiences create either fertile ground for trust or a glass walkway that is easily shattered. The experience also reinforced how important trust is in relationships. I strongly believe that people do not follow leaders they do not trust.

Trust is defined as "1a: <u>assured</u> reliance on the character, ability, strength, or truth of someone or something; **b:** one in which confidence is placed; **2a:** dependence on something future or <u>contingent</u>: **HOPE**" ('Trust,' 2024). Hope and reliance are two important words in this definition because I believe trust requires us to rely upon one another, and that trust is given with the hope that it will be reciprocated and deserved. I think of trust like a string between each of us. That string might be strong, like a thick piece of yarn that can withstand a bit of tugging and turmoil; or it may be weak, like a piece of tinsel that might break from the slightest gust of wind. I encourage you to think about the stakeholders you interact with by drawing yourself in a circle, with each individual or team of stakeholders as circles around you. Now think about the thickness of the line you would draw between yourself and each of these people or groups. Is it thin like tinsel? Is it thick like yarn? Is it maybe even a broken line where trust has been lost? This exercise can help you focus on which of your stakeholder relationships require more investment in trust.

I primarily use two sources in my work with people on developing trust, Stephen M.R. Covey's book *Speed of Trust* (Covey and Merrill, 2008) and the "Trust Equation" developed by David H. Maister, Charles H. Green, and Robert

M. Galford in their book *The Trusted Advisor* (Maister, Green, and Galford, 2021). Each of these frameworks contains elements that help define trust and how it is created.

Stephen M.R. Covey discusses 13 behaviors of high trust that I believe are a great starting point for how we think about trust and the behaviors we exhibit as leaders that build trust (Covey and Merrill, 2008):

1) **Talk Straight** – Be clear in your communication so that people understand what you are saying.
2) **Demonstrate Respect** – Respect is a cornerstone component of trust, and extending respect to others signals your trusting intentions.
3) **Create Transparency** – This is easy to say and difficult to do at times, given the speed of information flow.
4) **Right Wrongs** – We all make mistakes, and the important next step is to admit it and try to make it right.
5) **Show Loyalty** – People trust people who are loyal to them, and who stand up for them during good times and challenging times.
6) **Deliver Results** – People trust people they can count on to come through on actions and assignments.
7) **Get Better** – Learning and improving our skills lead to higher levels of trust in relationships.
8) **Confront Reality** – Optimism is a great trait, and we also need to be able to face adversity head on.
9) **Clarify Expectations** – People want to know what a good job looks like and trust that you will not change the rules.
10) **Practice Accountability** – Holding yourself accountable for your actions demonstrates trustworthiness to others.
11) **Listen First** – People want to be heard, and they will have greater trust in you if you take the time to understand their point of view.
12) **Keep Commitments** – People want to know you are capable of keeping your commitments to others; or, if not, able to explain why.
13) **Extend Trust** – I believe that taking the high road, trusting people until they give you a reason not to, is an important leadership skill.

Each of these behaviors helps you build that trusting relationship that is necessary to lead and influence.

As an engineer, I appreciate it when social scientists turn behaviors into equations! The Trust Equation (Maister, Green, and Galford, 2021) is:

$$\text{Trust} = (\text{Reliability} + \text{Capability} + \text{Transparency})/\text{Self-Orientation}$$

Reliability – If I ask you to do something, can I rely upon you to do it? A high number is high trust.

Capability – If I ask you to do something, are you capable of doing it? A high number is high trust.

Transparency – If I ask you to do something and you might miss a deadline, are you transparent about this or other challenges? A high number is high trust.

Self-Orientation – Do you do things because it will make you look good (i.e., you are in it for you) or because you want the team to succeed (as in, you are in it for the team)? A low number is high trust.

> **High trust using a scale of 1 to 10 might look like this: $(7 + 8 + 9)/2 = 12$.**
> This person is highly reliable, has excellent capabilities, is transparent about issues, and is in it for the team.
> **Low trust using a scale of 1 to 10 might look like this: $(5 + 9 + 4)/9 = 2$.**
> This person is not very reliable although they are highly capable, are not transparent about issues, and are in it for themselves versus the team.

The trust equation is a effective tool, as it allows you to examine the elements of trust to determine where a breakdown in your trust in someone (or their trust in you) is occurring. "Is it because I cannot rely on them? Or that I doubt their skills or intentions?" It is important to consider this equation from the viewpoint of the other person as well. "Does this person trust me as a leader? And if not, why? What steps might I take to repair that trust?"

Drawing from the two frameworks and my experience, I believe the key elements of trust are:

- **Agenda/Intent** – This is about **why** the leader or teammate takes action. Are they doing it because they care about you and the team? Are they doing it because it will help you and the team be successful? Or are they doing it because of the glory, accolades, and promotions they will receive? I worked for one person, and the entire team clearly understood that we were that person's steppingstone to the next promotion. It did not feel good, and we did not trust their intentions.
- **Transparency** – One day, I was sitting in a meeting with one of my largest and longest-term clients and found out from the client, with no warning, that we had missed a report deadline during the prior week. My head swiveled to the consultant sitting next to me who had missed the deadline—if looks could kill! This is not transparency. Transparency is raising your hand and saying, "I'm going to miss this deadline," or "I know I promised this by Wednesday, but I'm not going to make it," or "I need help."
- **Admitting Mistakes** – This is closely related to transparency. We all make mistakes. We need to own up to these mistakes, apologize, and try to make things

right. There have been a couple of instances in my career where I did not make things right after a mistake, or I did it after a long time had passed, and I regret it. Hurt feelings, feeling wronged, feeling betrayed—these can all contribute to not taking the steps we need to in order to make amends.

- **Caring** – People trust people they feel care about them. I say to the members of my team, "*You are a human being first and an employee second.*" I want to know about people's goals and aspirations and how I can help them achieve these, even if it means losing them as an employee. I want each person I interact with to know that I care about them.
- **Reliability** – Can I count on you? Do you come through in a pinch? If I ask you to do something, am I confident that you will get it done? If not, then back to transparency, am I confident that you will tell me? We trust people we can rely upon to meet the commitments they make to us.
- **Capability/Expertise/Knowledge** – As a leader, manager, or subject matter expert, do you have the technical skills to do the job? If you do not know something, do you have the courage to say, "I don't know, I'll get back to you," or approach your leader about further training? It is okay to not know everything. It is enough to know you need to ask for help.
- **Humility** – I believe people do not trust those who they feel think they have all the answers. No one has all the answers, and to be unable to admit that undermines trust. Our ability to be humble, ask for help, and acknowledge our mistakes builds trust in our relationships.

Actions That Build Trust

Building trust takes more than understanding the concepts described above. It also requires intentional action. I recently started using an exercise with groups to drive this home. I will divide them into groups of three and have them work through three role plays, during which one person is the Leader, one person is the Team Member, and one person is an Observer who provides feedback to the Leader after the role play.

In the first role play, the Team Member is new to the organization and has just joined the Leader's team. I ask the Leader to welcome them in a way that builds trust from the start. It is interesting how many people will start talking to the Team Member without giving any thought to what they are going to say! Think, then speak! What works well in this scenario to build trust? I suggest starting with great questions that show care:

- Ask them about their past experiences. What made them want to join this organization? How are they feeling about their first week?
- Tell them you care about them and their success at the organization, and that your job as their coach and leader is to ensure that they are successful (intent!).

- Share your expectations with them and how you will ensure that you are both on the same page. Share things like the best way to ask questions and how frequently you will meet.

The second role play is a bit more challenging. The Team Member is an effective employee who has moved to the Leader's team from another team where they had a difficult time. The Team Member has lost trust in the culture of the organization. Many more people stop to think and jot down notes before diving into this scenario, and there is typically tremendous debate on one point: Do you ignore the fact that they've had a difficult time or mention it? My thought is to ask great questions like, "Can you tell me about your experience on your prior team?" That leaves the door open for them to share what they feel comfortable sharing as you build trust. They might not feel enough psychological safety or trust to tell you anything, and that is okay. Reassure them that your role is to ensure their success and build trust over time.

The third role play is more challenging, and usually elicits a groan from the group. The Team Member has been a member of the Leader's team for quite some time and is an excellent employee, but has recently begun missing assignment deadlines. I ask the Leader to assume they have started to lose trust in this employee and to have a conversation that starts rebuilding their trust in each other. Observers typically report back on positive steps, such as Leaders who ask great questions ("You haven't seemed yourself, is everything okay?") and Leaders who stick to the facts of what assignments were missed and speak candidly about reliability, capability, and transparency. One important point of this role play is that trust is a two-way street. Leaders work hard to be trusted and must understand their level of trust in those they lead. Losing trust in an employee can lead to micromanagement and other unhelpful behaviors that continue to erode trust.

> Additional factors impact trust, including people's backgrounds, cultures, and ethnicities. Years ago, I was hired to work with an engineering team that was not performing well. When I first met with the co-lead, one of his initial comments to me was, "I'm from this small town in XYZ country, and we do not trust anyone. Just wanted to let you know." I appreciated him telling me this, and I believe it illustrates that our backgrounds also provide a lens as to how we approach trust.

I believe we sometimes extend trust without trust being returned. We need to realize that perhaps something has happened to that person that renders them incapable of trusting you, and that is okay. Everyone experiences a different version of you. During one experience, I worked with a person who had been taken advantage of in such a way that no one was going to earn their trust for a long time. It was hard for me to walk away from a trusting relationship with this person, but I had to realize this was about them, not me.

A breakdown in trust can feel like betrayal and be hard to recover from. I believe our ability to recover trust, to move past actions that deplete trust, and to rebuild trust in the absence of trust is an incredible leadership skill.

> Author and psychologist Beth Hedva shares this: *"When you're looking at how to shift your attention from the pain and the grip that betrayal has on you, you turn your attention inward and start by relaxing your body. I use the acronym TRUST. The 'T' stands for take a time out, turn inward to discover the deeper truth beyond the hurt, pain, and fear. The 'R' then says what you must do to let go of being gripped by that. You relax the body, start with long, slow deep breaths. According to the research of the Heart Math Institute it only takes two minutes of long slow deep breathing to change your entire neurochemistry from a stress response to a soothing oxytocin trust response, so take long, slow deep breaths, relax your body, release your mind. Then you do the 'U' in TRUST, use your intuition, use your body as a resource, your heart, recognize the energy in your thoughts. If your mind is spinning, ask your heart, 'What is it that is making my mind spin?' When those three things happen – body relaxed, heart open and full in a positive way, and the mind is calm and clear – you know you've hit the truth. That's when you speak your truth, the 'S' in TRUST, share your gifts. The final 'T' in TRUST is take action and includes turning inward and trying again because what often happens is when we start speaking our truth, guess what - some people don't like it."*
>
> (Amy Edmondson, speaking to Gagel, 2019 Greatness Podcast)

Building trusting relationships requires hard work and time, and can be unintentionally destroyed by one simple action. I believe great leaders reflect upon their level of trust in others and how much others trust them, and take deliberate steps to strengthen trust. People do not follow (and are not influenced by) people they do not trust. Investing in the trust level of your relationships is critical.

Activity:

- Think about those you lead and interact with at work. What is your level of trust in them? What is their level of trust in you?
- Think of one person within your organization that it is critical for you to trust and for them to trust you, where those levels of trust could perhaps be improved.
- What one thing might you do to build more trust with this individual?

Psychological Safety

During my career at Beech-Nut Baby Foods, I switched from leading the manufacturing operations to leading the distribution warehouse to broaden my experience. I once again found myself in a position where I knew little about the operations I was leading. During my first week, Art (a 35-year employee) asked if he could take me for a ride around the warehouse to discuss some things, and I said "yes." During the tour of this 250,000-ft warehouse, Art explained everything that was wrong with how it was organized and how that contributed to our poor load accuracy and productivity. For example, we were storing English label applesauce next to Spanish label applesauce, and drivers frequently picked a pallet of the wrong language for a load.

When we returned to the warehouse office, I suggested that we head up to engineering for a paper plot of the warehouse so that Art could draw up how he thought the warehouse should be organized and present his ideas to the three shifts of teams for input. I was shocked when Art said that during his long career, he had never been in the company's offices just adjacent to where we were standing. My response was, "It's about time you visited," and off we went to the engineering office. Art drew up a color-coded diagram of how he thought the warehouse should be organized to improve our team's performance and presented it to all three shifts. Each team added their ideas, and weeks later, the warehouse team reorganized the entire warehouse without one hour of overtime. Our load accuracy and productivity improved significantly.

Later, I asked Art if he had ever shared these ideas with other warehouse managers, and he indicated that no, he had not. I asked Art why he had shared these ideas with me, and I will never forget his response: "The manufacturing team said you would listen and be willing to try new things." What Art was describing was psychological safety, a team atmosphere where people feel free to bring up issues and ideas and to try new things without fear of being punished or humiliated (Edmondson, 2018). Unfortunately, research tells us that many people do not feel psychological safety at work. According to a recent Gallup poll, about 30% of employees strongly agree that their opinion is valued at work (Gallup, 2017). When people do not feel that their opinions are valued, they are hesitant to raise issues and the results can be disastrous. Examples such as the Boeing 737 Max, where employees knew there were issues, exist in nearly every industry (Koenig, 2020).

Interest in psychological safety was fueled by a *New York Times* article on Google's Project Aristotle that identified psychological safety as the most important factor in team performance (Duhigg, 2016). Luckily, author and Harvard

Professor Amy Edmondson had been researching the importance of psychological safety, having written her first paper on the topic back in 1999 (Edmondson, 1999).

> "*The most important aspects of framing the work are that it's challenging and uncertain and will therefore include missteps along the way. The second thing is to be proactive about inviting engagement. If you ask a subordinate or a peer an authentic question, a genuine question, and then you pause to hear what they have to say, you have created a little pocket of psychological safety for voice, especially if the question is asked in a thoughtful way. You would not want to use a bullying question or a leading question or a 'you better get the right answer or else' kind of question but a good, curious question. Then, we must listen intently. That creates psychological safety. And the third thing for creating psychological safety is to be thoughtful about how you reply. I'm not saying everything that everyone says is pure gold. It isn't. And you may find yourself thinking 'Well, that's a stupid idea', or 'I can't believe that's happening and you didn't tell me earlier'. In other words, you might have a knee-jerk response to bad news or to a crazy idea. But you can discipline yourself, whether you're a boss or a peer, to have a what I call a productive response instead, It is productive because we live in an uncertain complex world. None of us has all the answers at all times. A productive response is one that appreciates the effort or the risk it took to say something at all.*"
>
> (Amy Edmondson, speaking to Gagel, 2019 Greatness Podcast)

The first part of the psychological safety framework involves "setting the stage" to reach alignment on the important goals of the team, our key measures of success, and what's at stake if we don't achieve them (Edmondson, 2018). At Beech-Nut Baby Foods, our management team was told that we had 12 months to turn the company around and make a profit or Ralson Purina was going to shut the company down—a highly motivational context. The expectations were clear (make a profit), as were the stakes (shutting down the company). Amy Edmondson discovered team psychological safety while studying medical teams, where she learned that the higher-performing medical teams appeared to have a higher number of patient errors not because they were making more mistakes, but because they felt safe to report the mistakes and learn from them (Edmondson, 2018). For these medical teams, patient safety was their common goal, and "set the stage" for what was important.

The second part of the psychological safety framework is to practice humility and vulnerability by acknowledging that you do not know how to solve the problems, and to invite the participation of those closest to the problems to help solve them (Edmondson, 2018). That is what I was unknowingly doing as a young leader at Beech-Nut Baby Foods during my time leading both manufacturing and distribution. When I arrived, I knew nothing about making or distributing baby food. What I did know was that I had 800 people who knew how to solve our problems, and that if I walked the floor of the plant every day, and if I gained their trust and listened to them, these people would tell me how to fix our problems, and they did.

> *"The leader sets the tone for how the team is going to operate, and when they can open themselves up to being vulnerable and it being okay, then the team has permission to do that."*
>
> (Amy Edmondson, speaking to Gagel, 2019 Greatness Podcast)

We implemented what was a hot topic at the time ("self-directed work teams") to invite participation. This helped us shift from what Amy describes as a concept of bosses who know everything and give orders to bosses who respect employees as experts who can contribute to thinking on how things should be done (Edmondson, 2018). We set up teams of people to tackle all types of issues and come up with ideas on how to fix things, and it worked. *"Leaders being more vulnerable and being more human and approachable ... when you share a story that shows vulnerability, it creates that nice safe culture where it's okay to make mistakes and it's okay to not have all the answers"* (Gabrielle Dolan, speaking to Gagel, 2021 Greatness Podcast).

> *"Everybody in the room knows that there is a compulsion to sort of prove over and over again that you're in charge. Our view would be flip that on its head, to ask yourself, 'do I need to prove that I'm in charge?' No. 'Do I need to invite other people to share information that's going to make our team perform better?' That's more important than proving that I'm in charge."*
>
> (Edgar Schein and Peter Schein, speaking to Gagel, 2019 Greatness Podcast)

It also pays to learn people's craft because it shows you care and that you respect and value their expertise. I went to "retort school" to learn how retorts work, the cooking vessels that enabled us to keep baby food on the shelf for up to two years. I went to "glass school" to learn about glass defects that caused breakage. People admired how hard I worked to learn, and I never acted as though I knew more than anyone else.

The third stage of psychological safety then became critical at Beech-Nut—acting upon ideas and eliminating fear of failure. We tried many new ideas at Beech-Nut. Some worked, and some did not. If they did not work, we asked ourselves "why," learned, and moved on to the next iteration of the solution without ramifications for the failure. It took time to build the trust necessary to have true psychological safety, but I would say that within four-to-six months, it was strongly felt throughout the manufacturing facilities in both New York and California.

Interestingly, just before I left Beech-Nut Baby Foods to marry and settle in Denver, we hired a new leader from another manufacturing organization who became my manager. I am going to guess this person came from an organization with low levels of psychological safety. This leader had little trust in all of us crazy people running around listening to the plant employees and trying new things, and that caused a big disruption in our culture. This is a really important point. Changes in leaders, changes in team members, impact the dynamics of the team and the culture of the organization, and you sometimes have to start over and rebuild psychological safety.

Psychological safety is not about letting people off the hook or not holding them accountable for performance. It is not about everyone just being friendly and not focusing on results. *"I tell people very clearly, I hope, what psychological safety is and just as importantly what it isn't. It's not about being nice. It's not soft and fuzzy. It's about candor and directness and being willing to say it straight even when you're not 100% sure of how it will end up"* (Amy Edmondson, speaking to Gagel, 2019 Greatness Podcast). It is okay to build psychological safety AND hold people accountable for results. Just as with trust, your efforts to build psychological safety will require reflection, deliberate actions, and time.

Activity:

- Think about your team, department, or organization. What is the current level of psychological safety in this group?
- What one thing could you do to develop a higher level of psychological safety? What's the next right conversation to have with the team about psychological safety?

Conclusion

As I have recounted, our management consulting team worked with an energy company that was missing their capital budget by an average of 14% per year, and we took this number down to 1% in 18 months. I was leading a three-person consulting team for this project, and it requires a strong consulting team (not just a group of talented individuals) to enact this type of change within a client.

We trusted one another. We trusted each other's capabilities as consultants, and we could rely upon one another to do what we said we would do. If we were going to be late, we asked for help and communicated with the team. We were singularly focused on the same intent and agenda—to do great work for this client and help them solve their problem. As the leader of the team, I demonstrated situational humility by stating that I did not have all the answers (maybe none!), and I encouraged the participation of the consulting team in crafting our possible solutions. I also acted upon their ideas. It was an amazing experience that exemplified high levels of trust and psychological safety.

Taking the time to invest in building a trusting, psychologically safe environment has huge benefits in that people and teams perform better with lower levels of stress and higher levels of enjoyment. Trust and psychological safety are complicated. Digging into the elements of each can help you, as a leader and influencer, understand how to build, rebuild, and maintain trust and psychological safety within your team and organization.

References

Covey, S. M. R. and Merrill, R. R. (2008). *The speed of trust: The one thing that changes everything*. Washington, D.C.: Free Press.

Duhigg, C. (2016). 'What Google learned from its quest to build the perfect team,' *The New York Times*, 25 February. Available at: https://www.nytimes.com/2016/02/28/magazine/what-google-learned-from-its-quest-to-build-the-perfect-team.html [Accessed: 11 June 2024].

Edmondson, A. C. (1999). 'Psychological safety and learning behavior in teams,' *Administrative Science Quarterly*, 44, 250–282. Available at: https://www.researchgate.net/publication/313250589_Psychological_safety_and_learning_behavior_in_teams [Accessed: 11 June 2024]

Edmondson, A. C. (2018). *The fearless organization: Creating psychological safety in the workplace for learning, innovation, and growth*. Hoboken, NJ: John Wiley & Sons.

Gagel, G. and Dolan, G. (2021). *Gabrielle Dolan discusses her book Magnetic Stories* [Podcast]. 2 February. Available at: https://open.spotify.com/episode/0uamofsvN18F1sWcjawMQm [Accessed: 28 April 2024].

Gagel, G. and Edmondson, A. (2019). *Amy Edmondson discusses her book The Fearless Organization on psychological safety* [Podcast]. 30 October. Available at: https://open.spotify.com/episode/6joytIVhV3XlGFZx6rcodC [Accessed: 28 April 2024].

Gagel, G. and Hedva, B. (2024). *Dr. Beth Hedva discusses her book Betrayal, Trust and Forgiveness* [Podcast]. 5 April. Available at: https://open.spotify.com/episode/7nMb4TNnXhfeD5p88Wkqco [Accessed: 2 May 2024].

Gagel, G., Schein, E., and Schein, P. (2019). *Ed & Peter Schein discuss their book Humble Leadership* [Podcast]. 24 June. Available at: https://open.spotify.com/episode/5JueFC2LHdo74Ncz4HqVs5 [Accessed: 28 April 2024].

Gagel, G., Terkelsen, J., and Terkelsen, M. (2020). *Jan and Michelle Terkelsen discuss creating high performance teams* [Podcast]. 3 February. Available at: https://open.spotify.com/episode/4U22Jcl7lYP2Ana37I4zrI [Accessed: 28 April 2024].

Gallup. (2017). *State of the American workplace.* Available at: https://qualityincentivecompany.com/wp-content/uploads/2017/02/SOAW-2017.pdf [Accessed: 11 June 2024].

Koenig, D. (2020). 'Messages show Boeing employees slid 737 Max problems past FAA,' *PBS News Hour*, 10 January. Available at: https://www.pbs.org/newshour/nation/messages-show-boeing-employees-slid-737-max-problems-past-faa [Accessed: 11 June 2024].

Maister, D. H., Green, C. H., and Galford, R. M. (2021). *The trusted advisor.* 20th anniversary edn. Washington, D.C.: Free Press.

Outward Bound. (n.d.). *Who is Outward Bound?* Available at: https://www.outwardbound.org/ [Accessed: 11 June 2024].

'Trust.' (2024). *Merriam-Webster Online Dictionary.* Available at: https://www.merriam-webster.com/dictionary/trust [Accessed: 11 June 2024].

14

The Three "C's" – Communication, Collaboration, and Conflict Resolution

Introduction

Years ago, one of our investment bankers was attempting to work with the leaders of two newly acquired companies to accomplish a brand merger and targeted acquisition. There was such a lack of communication and collaboration, such a great deal of conflict (and ensuing bad behavior) within the newly combined executive leadership team, that they were making little headway. I liked and admired the CEO of the two merging companies and knew he genuinely needed help, so when the investment banker reached out to me for assistance, I agreed. We arranged a two-day meeting with the 12 executive leaders to begin working through their issues, and I specifically asked that we meet for an hour the night before our meeting and then attend a group dinner. This was met with grumbling about cutting their golf game short, but they acquiesced.

Of course, my plane was late, and I arrived to find 12 men sitting around a rectangular meeting table, staring at me, practically daring me to help them resolve their issues. After some introductory remarks to build my credibility, I asked each person in turn to share their DiSC profile, and I was not surprised to find that 11 of the 12 men had "D" as either their first- or second-highest element. "D"'s love to be in control, and my first suggestion to the group was to come to grips with the fact that 11 of them could not be in control of everything, all the time!

I then went around the table and asked every man to rate himself on a scale of 1–10 on how savvy he was as a businessperson. Each rated himself 8–9, with comments like *"Well, I guess I'm not perfect,"* but they were all definitely feeling confident in their abilities as businessmen. I then asked the group, "On a scale of 1 to 10, how professional are you as a group?" This question was met with some reservation, and they could see where I was headed. I did not ask each of them to answer this question. Instead, I waited until one or two volunteered, *"Well, I guess we're fairly professional,"* when in fact I knew that some of their behaviors such

Building Women Leaders: A Blueprint for Women Thriving in Construction,
First Edition. Gretchen Gagel.
© 2025 John Wiley & Sons, Inc. Published 2025 by John Wiley & Sons, Inc.

as storming out of meetings were not professional. But I had them. I said, "*Okay, I am going to assume that you all are savvy businessmen who act in a professional manner and who are smart enough to figure out how to merge these companies and solve any challenges we might face in the process. Let's go eat!*"

Over the next two days, we made a list of every challenge we needed to solve, every strategy we needed to put in place, and we worked together—professionally—to create the necessary strategic and tactical plans to enact the merger and plan for a future acquisition. A year later, incredible progress had been made and it was a high-functioning executive team.

I call these three topics the "Three Big C's" because I believe that a leader's skills related to communication, collaboration, and conflict resolution are critical to building strong relationships. Many times, I have come into a sticky situation on a project or within an organization where there is a significant lack of skill in one of these areas. It is often not pretty, and it can wreak havoc and cause incredible stress.

> I want to add a point about maturity here. I walk into every meeting assuming everyone is a mature adult. For some people, that is probably a stretch, but it is my assumption. I once had a female engineer I was coaching say to me, "*You're reminding me to be an adult!*" I can go to drama as fast as the next person, but the older I become, the less I want drama and the more I realize the value of emotional maturity. Approaching these three topics with maturity is critical. We as leaders need to take the high road, employ emotional intelligence, and stick to the facts while being aware of our assumptions.

As you read through this chapter, reflect upon your skills in these areas. Where are you strong? Where might you improve? What specific tips will you take from this chapter? Effective communication, collaboration, and conflict resolution pave the way for outstanding results and less stress in your relationships.

Communication

You as the "Sender"

Have you ever played the game "Telephone," where people sit in a circle and one person whispers a phrase into the next person's ear, with the process repeating for the entire circle? It is such a simple exercise, and yet rarely does the phrase uttered by the last person sound at all like the original phrase. One year, I was conducting a performance review for a construction industry CEO, and one theme that emerged was that his team did not understand his vision of where the organization was headed. When I shared this feedback, he nearly erupted. "*I communicate my vision constantly!*" He felt so, but his team did not. Even when leaders feel that they are overcommunicating a message, that communication effort can fail.

In every communication, there is a message sender and a message receiver, and both play an important role in ensuring that communication is clear and productive. Each party has a responsibility in every exchange and conversation, and every interaction is an opportunity for brilliance or failure (Connolly and Rianoshek, 2002). Your responsibility as a leader when "sending" a message is to make sure that: (1) the message is as clear and concise as possible (KISS—Keep it Simple, Stupid); (2) distractions such as cell phones are not present while you are communicating; (3) the message is received correctly by the "receiver;" and (4) the message is repeated often when communicating to teams.

> "*I think the mistake many leaders make when they're trying to communicate is they feel like they must communicate everything, and the reality is people can't take in everything. What is your one key message? If you think of how you can get your message across in the shortest amount of time possible it forces you to be really concise and be really clear on what your message actually is. You probably need to know what your three main messages are and too many people say, 'But I've got 10 very important messages', and when you try to get 10 across, people end up remembering none of it because it's too overwhelming and too confusing.*"
>
> (Gabrielle Dolan, speaking to Gagel, 2019 Greatness Podcast)

Let's say you are sitting down with someone to communicate a major change in a project schedule. Prior to the conversation, you might jot down three bullet points of the most critical information you want to convey. You might then ensure you have their attention and that they are not distracted. You might say, "*This is really important information, is now a good time to talk?*" After you have shared the information, you might ask them to repeat the key points back to you. You might also follow-up with an email confirming the three key points of information and ask them if they have any questions.

> "*At the end of the day it's not about what you say as a presenter, it's about what they receive as an audience. The point not only makes you, the speaker, clearer, but makes it easier for your audience to receive it, reflect on it, and attach their relevance to it. We think we know what a point is because we hear that word all the time – 'get to the point.' Where we make our mistake is confusing points for other things like topics, themes, categories, catchphrases, and observations. A true point is a proposal you are making to your audience – if we do X, Y will result.*

> "*I have a test to help people understand if they have a point or not. It's called the 'I believe that' test and it goes like this. You take what you think is your point, put the words 'I believe that' in front of it, and ask yourself one question: 'Do I now have a complete sentence?' Not a run-on and not a fragment. If it is a complete sentence, you're on your way to making a point.*"
>
> (Joel Schwartzberg, speaking to Gagel, 2023 Greatness Podcast)

Crafting your communication in a thoughtful way requires time. Running from meeting to meeting is not helpful, in that it sets us up for communication failures that are painful and costly because we do not have time to prepare to communicate. Carving out time to think before you speak, before you present, before you interact, can pay big dividends in your efforts to lead and influence. Taking the time to prepare for your communication, to be clear on the point you are trying to make and de-clutter your communication, are highly rewarding in that people appreciate clear communication. You control how you deliver the message, and great leaders are adept at communicating with people in a way that creates connection and clarity.

Communicating with Teams of People   What does effective communication look like in our day-to-day interactions with teams of people where the dynamic within the team might further complicate our ability to communicate? Let's say there is a change coming down from above, and that a new time reporting system is being put into place for all construction projects starting next month. You might call a meeting with your team, and the day before the meeting, you might sit down for a few minutes and think about:

1) **What tone am I using to convey the message?** Positive is a good start, and throwing those above you under the bus is rarely effective, even if you do not agree with the change.
2) **What are the two key sentences that I want to make sure I repeat and key points?** For example:
   a) Starting June 1, we are going to be implementing a new mandatory time reporting system on all construction projects.
   b) I realize that it may be unclear as to why we need this system (showing empathy), but here's why this system is important.
   c) When we put a bid together for a job, we are making certain assumptions about how that project team will spend their time. Sometimes, these assumptions are wrong. If we collect good data, we can feed that information back into the estimating and bidding process and hopefully win more jobs, but also have projects that are more profitable.

d) The more profitable our jobs are, the more profitable the company is, and we all benefit (the WIFM, "what's in it for me") through… bonuses, ESOP ownership, job security, etc.

e) I may not be able to answer all of your questions right now, and if I can't, I'll find out the information and get back to you.

f) Again, this new time reporting system is mandatory. Although I realize it will take some getting used to, it will provide invaluable information to help our projects and company be more profitable.

g) Questions?

3) **What are the questions that people are likely to ask, and what are my answers?**

h) What happens if someone doesn't use the system? My answer is…

i) How is the information about time going to be used? Stored? My answer is…

4) **What fears might people have, and how might I allay those fears?**

j) There may be people on your team who cannot read, as we found out with one of our university graduates at Ralston Purina. It's not as uncommon as people think.

k) People may be afraid they cannot learn the technology. You might add a statement like, "We are here to help everyone be successful in using the technology through training, etc."

l) People may feel they will not have enough time to input their time.

I suggest following up with an email to everyone who attended and those who may have missed the meeting that lays out what was said and the questions, along with their answers. I might also follow-up specifically with those who missed the meeting to see if they have any questions. Is all of this time-consuming? Absolutely. But I find that people rarely spend the time they need to prepare for communicating important messages to teams, and this upfront and follow-up time could save you hours of heartache later.

Diagnosing and improving communication breakdowns form a vital leadership skill, and focusing on yourself as the "sender" of the message is an important starting point. Communication is a huge topic, and I have only skimmed the surface, focusing primarily on oral communication, although the ability to communicate in writing is also an important skill. I encourage you to continue learning about ways to effectively communicate and to help others do so as well.

Activity:

- Reflect on your personal communication skills. Where could you improve? What communication skill might you further develop?

- How might you carve out more time to reflect before communicating with people and teams?
- Think about how well your team communicates. What type of communication training might be useful to your team? How can you, as an informal and formal leader, help the team communicate more effectively?

Understanding the Receiver

Early in my consulting career, I was working with a contractor team to document their project control process. I thought, "How hard could this be?" Really hard! Nearly everyone had a different opinion of how project controls should be done. But that was not my greatest learning experience on this project. The co-owner of this construction company was a big-picture thinker, which was one of the reasons he had hired me to work with his team to dig into the details of project controls—he didn't want to do it. I was a young management consultant and my go-to was to flood him with information. Bad idea. We worked it out in the end, but the project would have gone much more smoothly if I had taken the time to consider my audience when communicating with this client.

> "*Often in business we communicate something, and we think we've communicated, and then people are saying, 'what new strategy', and you say, 'We had the whole presentation'. We think we've communicated, but if they don't get it, it's not their fault. It's our challenge to figure out how to help them understand it.*"
>
> (Gabrielle Dolan, speaking to Gagel, 2019 Greatness Podcast)

Taking the time to understand the unique characteristics of those we are trying to communicate with is at times daunting, but also critical to effective communication. I joke that leadership would be easy if everyone was like the Stormtroopers coming off the transports in *Star Wars,* identical and predictable. I am a big fan of using profiles such as the DiSC profile to understand people and the style of communication that works best for each individual (DiSC Profile, n.d.). The DiSC profile helps you tailor your communication to meet diverse "receiver" needs; to understand who wants the six-sentence email, who wants the detailed email, and who needs data, for example.

This is not about being disingenuous or manipulative. This is about stopping to think about the needs of the "receiver" and crafting your message in a way that meets those needs. Are they a "D"? Get to the point, six bullets and done! Are they an "I"? Start with small talk, ask them about their weekend. Are they an "S"?

Explain the logic of your thinking. Are they a "C"? Give them the details, all the details! It is not necessary to have people complete the DiSC profile to understand your audience; you just need to invest time thinking about their preferences and how they communicate. Do they show impatience? Do they often ask for more details? We also need to understand that we frequently communicate in our preferred style instead of the style of others. As a "D" and "I," I like to be direct and skim over the details, and that does not work for a large portion of my audience. It takes effort for me to adjust, but it is worth the investment of time.

> When I was CEO/President of WFCO, I secured a significant donation from a first-time donor who I knew could give more if we delivered on the promised results. I also understood from meeting with this person that they were a "D" and "I" DiSC profile—direct, to the point, and a relationship builder. Whenever I sent this person an update email about their gift, it was short, newsy, and focused on the results we were achieving with the money, ten sentences long at maximum with a picture and an offer of more information if necessary. We developed a great relationship, and I did in fact secure a much larger second gift from this donor. I took the time to think about this person and their communication preferences, and I believe that demonstrates respect for others.

Selecting the correct communication channel for your audience and message is also important. When I started meeting with the executive team I described earlier in this chapter, I realized that the two groups of executives, located in two different cities, were adept at firing inflammatory emails at one another. We created one essential norm early in our discussions: if someone had a problem with someone else, an action that frustrated them, they were not allowed to email that person about the problem, but instead had to pick up the phone. I made them promise. When we met again three months later, the change was astonishing. They had kept their promise, and instead of firing inflammatory emails at one another, they were talking and resolving issues. I was quite proud of them.

We all have communication channels we overuse. I overuse email because I am consistently operating across multiple continents and time zones. Emails are appropriate in many instances, but not when you have a delicate topic to discuss, when you fear your tone will be misrepresented, or when it is critical that the emotion you are attempting to convey is understood. Choosing the right communication channel (such as an email, a text, a phone conversation, or an in-person conversation) is critical in building strong relationships.

It takes two to tango: a sender and a receiver. Investing time in considering your audience and their preferred communication style, taking time to ensure they are ready to receive the message, are important as a leader.

Activity:

- Before each meeting next week, take three minutes to think about the person/people you are meeting with and their needs as the audience.
- How might you adjust your communication style to better meet their needs?
- What communication channel do you perhaps overuse? Underutilize?
- What one step might you take next week to be more thoughtful in selecting the appropriate communication channel?

Speaking in Public

Sometimes, building relationships with people involves speaking in front of groups. I learned to speak to groups the hard way: standing up in front of teams of manufacturing employees (mostly men) and trying to convince them to do things in a different way. It was hard. I would encourage you to practice speaking in front of people because it's going to happen—in front of your team, a project team, or some group of people. I was a graduate teaching assistant for oral and written communication during my MBA program at the University of Denver, and I loved working with people who had an overt fear of speaking in front of others. Many, many people suffer from this fear, and overcoming it is a powerful step in improving your ability to communicate with others.

> *"When I work with women, why do people shrink back from public speaking? Where do those fears come from? There's a voice in your head that throughout your life has told you, look both ways before you cross and maybe don't eat that. But in public speaking that voice becomes the spokesperson for your insecurity. It is lying to you, and I can't tell you how many people, both men and women, will give a perfectly effective speech and then stop and say, 'Oh, that wasn't good.' I give that voice in your head the name Roy, and I say, 'Roy is really trying to sabotage your presentation.' If Roy is strong enough, not only will he make you think they're not liking it, it's boring, this is not going to be interesting; but he will take over your mouth, and that's when you have speakers say things like, 'Well, I'm not really prepared today but all right', 'Well, this may not be the best but all right', 'Well, I don't know if you like this or not.' But that's not them. That's the voice in their head who's so powerful that he actually came through their mouths, and they might be lies. We think they're truth because they're coming from ourselves. But in truth they might not be consistent with what's actually going on."*
>
> (Joel Schwartzberg, speaking to Gagel, 2023 Greatness Podcast)

Here are a few of my tips for speaking in front of groups:

- **Find friendly faces in the audience and focus upon these people.** You might even ask them to be encouraging ahead of time. If I am speaking to a large audience, I look for a friendly face on the left and a friendly face on the right. Those are the only two people I need to look at, and everyone thinks I'm looking at the entire audience.
- **If possible, prepare but don't read your remarks.** I start by writing out the first three-to-four sentences to start me down the right path, then I move to bullets and write out full transition sentences to effectively guide myself and the audience through topic changes.
- **If you are not nervous, you should be!** Nervous adrenaline is a good thing, as it keeps you on your toes and gives your speaking energy. Calm your nerves through deep breathing and practice. Envision yourself giving an amazing speech the same way athletes use positive visioning. Think to yourself, "I am going to crush this!"
- **Practice avoiding filled pauses—saying "um," "you know," etc.—ahead of time**. I hear myself using filled pauses on my podcasts and I cringe because I know how to fix it and I forget! I call these "filled pauses" because your brain is pausing (thank goodness, it needs to think), and you "fill" that pause with one of these words or phrases. While you practice your remarks in front of a mirror, practice silently pausing for three-to-five seconds throughout. It will feel uncomfortable at first, but it works. We all need to pause to gather our thoughts while we are speaking, and through practice, we can avoid filling these pauses.
- **Practice, practice, practice.** Signing up for a course in public speaking (such as Toastmasters) can also be helpful.

As we discussed with personal branding, storytelling is a powerful way to convey your message because stories paint a vivid picture of the point you are trying to convey and create an emotional connection with your audience. When I stand in front of groups and share the story of my father telling me I should not go to engineering school and how that fueled my passion for more inclusive thinking, that story connects.

> *"If you're in a situation where you've got to deliver a message and you want that message to be really understood and remembered, then a story is a powerful way to do that. If you're giving a presentation I say, if they're only going to remember one message, what do you want it to be, and have a personal story for that."*
>
> (Gabrielle Dolan, speaking to Gagel, 2019 Greatness Podcast)

The ability to speak in front of people supports your efforts as a leader to influence the behavior of others. Investing in this skill is important, and if this is an area of fear for you, I am confident you can overcome it.

> **Activity:**
>
> - Write down three strengths you have when speaking in front of teams and groups of people.
> - In what one area might you improve? What tip from this section might you practice?
> - What story might help you connect with your audience the next time you speak in a group?

Collaboration

For decades, I have used a game called "Win as Much as You Can" with organizations' leadership teams and cross-organizational construction project teams to stress the concept of collaboration and how lack of alignment with common goals can sink organizations and project teams (Harvard Law School, n.d.). In the game, a group of people is divided into four groups and only provided with these instructions:

> 1) You are trying to "win" as much as you can. If they ask who "you" is, I just repeat this statement.
> 2) I provide the voting chart found in Table 14.1.
> 3) In the first round of voting, I provide each group one minute to decide their vote—schedule pressure!
> 4) Invariably, the votes are a mix of X's and Y's, and I record the score for that round, with some groups gaining points and some losing.
> 5) After Round Four of voting, I ask each team to send a representative out into the hall to discuss how it's going and determine the next vote. Sometimes, this group of representatives will agree that everyone is going to vote "Y," so that every team receives points. However, frequently, when the representatives return to their groups, they disagree with the decision and override the agreement. That starts some fireworks in the room!

This repeats for 10 rounds, with the dynamics often replicating the dynamics in an organization or on a project team where different departments or subcontractors are keeping their own score and fighting for resources, often with detrimental results. By the end of the exercise, the groups realize that the "you" in this game (and in organizations and on construction projects) is ALL of us. Not a subgroup of us, all of us. They realize that if they had all voted "Y" in each round, everyone would have won more.

Table 14.1 Win as Much as You Can Voting Payoff Schedule.

All Four Teams Vote "X"	Every Team Loses $1
Three Teams Vote "X" and One Team Votes "Y"	The Teams Voting "X" Win $1
	The Team Voting "Y" Loses $3
Two Teams Vote "X" and Two Teams Vote "Y"	The Teams Voting "X" Win $2
	The Teams Voting "Y" Lose $2
One Team Votes "X" and Three Teams Vote "Y"	The Team Voting "X" Wins $3
	The Teams Voting "Y" Lose $1
All Teams Vote "Y"	All Teams Win $1

Collaboration is defined as **"1:** to work jointly with others or together especially in an intellectual endeavor; **3:** to cooperate with an agency or instrumentality with which one is not immediately connected" ('Collaborate,' 2024). Collaboration abounds when people take time to explore and understand what the common goal is and how collaboration will allow each person, team, and organization to "win" together. People must understand that the whole is greater than the sum of the parts.

Harvard professor, author, and speaker Francesca Gino shares: "*Collaboration is not a value to impart but really a skill that needs to be developed. I spent time at Second City trying to understand how they use principles from their improv comedy training to help collaborative efforts and a few ideas stood out to me. One is active listening. This is something that I think sounds intuitive but think about the last meeting you were part of. There is usually a desire to speak, sometimes people dominate the conversation, or even when they think they're listening, they're just getting ready to talk. What's interesting is that improv comedy is all about your ability to carefully listen and give each other space before you contribute to the scene.*

"*One of the things that I often tell my students or leaders when I teach these skills is that curiosity and judgment cannot coexist. Again, this is a principle that very much comes from improv comedy but it's an important one, that if we are listening to the ideas that others are bringing to the table or to their arguments, there is a desire to judge. We go very quickly to 'that's not a good idea' or maybe 'it's not as good as ours.' But if we're able to put that judgment on hold and truly come to the table with curious eyes and wanting to understand the other point of view better, we are going to be in for a much more productive conversation and as a result of that a much more productive and easier and effective collaboration.*

> *"Groups tend to work best when we are making choices that are very conscious, choices about when to lead and when to follow. What that means is that you might in fact be the person at some point convincing people that there is a right course of action or decisions that need to be made. But you're also the person who's willing to sit back and let others take the lead. Take, for example, jazz bands. Often what you see jazz players do quite well is to have this willingness to lead and also to follow. When we're able to collaborate like that in our business we fare better."*
>
> (Francesca Gino, speaking to Gagel, 2020 Greatness Podcast)

Collaboration is not just a personal relationship challenge but also challenging within a team, department, or organization. I appreciate a question Patrick Lencioni poses in his book *The Advantage*: does your team or organization keep score as a basketball team or as a golf foursome? (Lencioni, 2012). A golf foursome has a collegial time playing golf together, and may even congratulate each other on good shots, but at the end of the day, every person is posting their own score. A basketball team, on the other hand, only has one score, and everyone contributes to achieving that score. I was coaching the CEO of an organization whose team was not happy with him. I realized in our conversations that he was seen as the only person putting points on the board, and his team did not understand how they put points on the board. How you keep score on your objectives drives culture and behaviors within your organization.

I feel as though the ability to collaborate is one of the most important skills in our industry, and while we have just scratched the surface of this topic, I hope this has inspired your thinking and fueled your curiosity to learn more. It is important to think about how you as an individual can more effectively collaborate, as well as about how you as a leader can help those around you collaborate.

Activity:

- Is your team/business unit/department a golf foursome or a basketball team? Is this driving behaviors consistent with the desired culture and values of the organization?
- What would great collaboration look like for your team? How could it benefit your team?
- What might you do as a leader to reward collaboration?

Conflict Resolution/Negotiation

There is a story in the book *Getting to Yes* that I have probably shared hundreds of times (Fisher, Ury, and Patton, 2011). Two chefs are in a kitchen preparing a meal, and they both reach for the only orange in the kitchen to use in their recipe. Both chefs immediately become positional, and an argument ensues. "*I need the orange.*" "*No, I need the orange.*" The chefs do what most of us do in these situations: they cut the orange in half, and neither is fully satisfied. As they are preparing their meals, they look at one another and realize that one chef needed the rind of the orange while the other needed the inner segments of the orange. Had they paused to explore their needs and expansively explore solutions, they would have realized their differing needs and come to a solution that fully satisfied both.

> "*We often talk about the importance of moving from positions to interest. It's not the 'what' but is the 'why' you want, and if you're able to go to the interest, you're going to uncover aspects that are not as different from what the other side wants.*"
>
> (Francesca Gino, speaking to Gagel, 2020 Greatness Podcast)

How we resolve conflicts with individuals is critical in maintaining positive relationships. As this example illustrates, the first important step in resolving conflict is for each stakeholder to clarify their needs and then think expansively about how we reach a solution that meets all the stakeholders' needs (Fisher, Ury, and Patton, 2011). I created an acronym, "FST," to think about the steps one might take: "Facts First, Solutions Second, Try Things Third."

Let's use a construction project as an example. Three subcontractors are all arguing about the laydown area they need, each saying they need more than they've been allocated. Here are the steps I would take:

- **Facts First** – It is important to focus on the facts versus assumptions and opinions. What are the facts of the situation? Who needs what laydown area, and for what timeframe? Is there any flexibility in this? The needs of all stakeholders should be described in detail to help everyone understand what is needed, and why.
- **Solutions Second** – It is important to interact with all stakeholders (together or individually) to brainstorm all the possible options available to solve this problem in a way that meets everyone's needs. This brings transparency to the process. It is important to encourage creative, even "crazy," ideas that may not have been considered in the past.

- **Try Things Third** – It is important to implement the solution and adjust as necessary. The solution may not work perfectly, but with everyone's focus on our common needs, we can keep tinkering with the solution until it works. We should also feed that information back into the system as a feedback loop for similar challenges we will face in the future.

> I experienced a great example of this process during the construction of a brown-field manufacturing plant addition. This company had built a similar project in Asia the previous year in nine months, and the current project schedule for this US site was 13 months and not acceptable to the client's internal marketing team. The client pulled together the major engineering and construction firms involved in the project for a collaborative design session and started with the facts, including the current schedule and relevant schedule components as well as the client-driven targeted end date. The team then used Lean construction methodologies to brainstorm possible solutions that would reduce the project schedule. Interestingly, it was the steel erection contractor who came up with a game-changing idea of resequencing the project that cut the project schedule by four months. The team was willing to "try things," and it worked.

One huge hurdle to overcome when applying collaborative thinking to conflict resolution is the tendency to jump to a position and really dig in your heels, like the chefs did with the orange. It is easy to think, "I need this," and stop at that thinking without being receptive to other possibilities. Your skill in overcoming positional thinking as a leader is an important step when resolving conflict.

> *"What's interesting is that often, especially if you and I are having a discussion and you suggest a view that is different from mine, the natural response is that I'm going to try to defend my view and fight you. As it turns out, that's not the best approach if we want to resolve the conflict quickly and have a productive interaction or a productive collaboration. My colleagues and I talk about the idea of a conversation of receptiveness. What we've done in our research is look at features of language that signal more receptivity, language that involves acknowledging the point that the other side is making or thanking them for bringing that point forward, and then making sure that you also state your view in clear terms. What's interesting is that when we show up with more receptiveness and we signal that to others, we end up being more persuasive."*
>
> (Francesca Gino, speaking to Gagel, 2020 Greatness Podcast)

Salary and Benefit Negotiations

At one point in my career, I felt I was not being paid the same as my male peers. I approached my boss and asked, "What are our salaries based upon?" Their response was, "Tenure and responsibility." I then pointed out that my tenure was exactly between two male counterparts, and that the three of us all had the same responsibility, so I assumed my salary was between theirs. After a significant pause, I was told he would get back to me, and I received a substantial raise. I hated that I had to raise the point. Why wasn't I making as much as them just because it was the right thing? Probably because I had not effectively negotiated my worth.

The ability to negotiate your salary and benefits is an important skill, and unfortunately, many of us avoid these conversations. I encourage you to start by knowing what the market rate is for your position and understanding the details of your compensation system within your organization. If you feel that you are not receiving adequate compensation or benefits, ask your direct supervisor for a conversation. Simply state, "I'd like to carve out 30 minutes to discuss my compensation. What is a good time for you?" If they ask why, say, "I just want to do a check-in to make sure I'm being fairly compensated for my role."

Come to the meeting prepared with the market and internal data, along with a list of your responsibilities and achievements. This is not the time to be humble. You need to be able to articulate the value you provide to your team. Start by asking about your compensation and how it compares to others in your peer group or the market. Depending on the answer, be prepared to state your case for why your salary and benefits should be higher. Stick to the facts and leave your emotions at the door. This is a factual conversation about your compensation within a compensation system and relative to your peers. If you do not receive an appropriate answer, or you do not feel your compensation is appropriate, state that you'd like to include human resources in a future conversation to resolve what is a difference of opinion about your worth to the team and the organization. This is important, whether you are male, female, orange, or blue. Do it. Have the courage to ask for what you are worth.

At the last meeting of the first cohort of the APGA Women's Leadership Development Program, I asked participants to share success stories from their time in the program. One participant stood up and said that she had asked for a conversation with her boss and shared with him the value statements of her worth that we had crafted during the initial session of the program. She stated that she did not even have to ask for the large raise she received because her boss then understood all the different ways she was contributing to the organization. Well done!

I joke that nearly everything in life is a negotiation, from where you are going to take the family vacation, to asking your kids to eat their vegetables, to convincing an employee to put on their safety gear. Leaders can reduce conflict within teams by effectively applying conflict resolution skills and teaching those skills to the members of their team. I encourage you to approach conflict resolution and negotiation with emotional maturity, use the FST model—Facts First, Solutions Second, Try Things Third—and use a growth mindset to explore many possible solutions before implementing the solution that meets the most possible needs of the stakeholders.

Activity:

- Think of a recent negotiation you had with a fellow employee. What went well? What did not go well? What would you have done differently?
- When you think of "Facts First," "Solutions Second," and "Try Things Third," where do you excel? What might you try to do differently in future negotiations and while working to resolve conflicts?

Conclusion

The ability to communicate, collaborate, and resolve conflict (my three big "C's") is vitally important to your ability to lead, and there is one "norm" I believe is critical to these three topics. If you have a problem with someone, a breakdown in communication, lack of collaboration, or conflict—talk to them (not 10 other people in your organization) about it. When I became CEO/President of The Women's Foundation of Colorado, I realized there were probably only about two degrees of separation between any two people in Denver. I made a vow that has withstood the test of time. If I don't want it to be a headline in *The Denver Post*, then I don't say it, text it, or email it. Period. That adage, "If you don't have anything nice to say, don't say anything at all," is true. Talking about people behind their backs will only harm your professional reputation. It will not improve communication, collaboration, or conflict resolution.

Communication, collaboration, and conflict resolution are three big topics that are critical to your success as a leader and influencer. Hopefully, this chapter has raised your awareness with a few great tips as a start and fueled your appetite to learn more.

References

'Collaborate.' (2024). *Merriam-Webster Online Dictionary.* Available at: https://www
.merriam-webster.com/dictionary/collaborate [Accessed: 11 June 2024].

Connolly, M. and Rianoshek, R. (2002). *The communication catalyst.* Fort Lauderdale:
Kaplan Publishing.

DiSC Profile. (n.d.). *What is DiSC?* Available at: https://www.discprofile.com/what-
is-disc [Accessed: 11 June 2024].

Fisher, R., Ury, W. L., and Patton, B. (2011). *Getting to yes: Negotiating agreement
without giving in.* London: Penguin Books.

Gagel, G. and Dolan, G. (2019). *Gabrielle Dolan discusses her book Real
Communication* [Podcast]. 21 May. Available at: https://open.spotify.com/episode/
2By0s6SjK4tfZpXmXYB4GC [Accessed: 28 April 2024].

Gagel, G. and Gino, F. (2020). *Francesca Gino discusses how to achieve high levels of
collaboration and her book Rebel Talent* [Podcast]. 24 July. Available at: https://open
.spotify.com/episode/3vPQUTRwjWtQFpAmwIyYNl [Accessed: 28 April 2024].

Gagel, G. and Schwartzberg, J. (2023). *Joel Schwartzberg discusses his book Get to the
Point* [Podcast]. 16 June. Available at: https://open.spotify.com/episode/
08Zpsyp0J1PZ4sKvU60w58 [Accessed: 2 May 2024].

Harvard Law School. (n.d.). *Teaching materials and publications: Business and
commercial role play: Win as much as you can.* Available at: https://www.pon
.harvard.edu/shop/win-as-much-as-you-can/ [Accessed: 11 June 2024].

Lencioni, P. (2012). *The advantage: Why organizational health trumps everything else
in business.* San Francisco: Jossey-Bass.

15

Clarifying Expectations and Providing Feedback

Introduction

One thing I miss when I am not in the US is live National League Football (NFL)—especially if it involves my hometown Kansas City Chiefs—and I will often watch a compilation of the game highlights online. One week, the big news was that tight end Travis Kelce was about to set a record for his eighth consecutive season of over 1,000 yards of receiving. However, when I watched the game highlights, one of the most excited statements from a commentator was something like, "Look at that block Kelce just made on that play!" Playing the position of tight end in football requires many critical skills, including catching, running, and blocking. The same can be said for professional golfers who have a long game (drives), a short game (irons), a sand game (that's me, in the bunker!), and a putting game. All these games must be working well to achieve the highest level of achievement, and professional athletes review statistics, practice their skills, watch videos of their performance, and receive lessons and coaching to assess their skills and improve.

I argue that we should treat the people on our teams like professional athletes, whereby we clearly define the critical skills of their role, regularly assess their performance, and provide coaching and training to help them improve. Many organizations start this process by creating a skills inventory or role framework for various positions. For example, you might list the skills required for an effective project manager to be high-performing in managing a construction site. But do most people in the construction industry clearly understand the expectations of their job? How effective are we at assessing people's performance relative to the skills that are on this list and providing consistent coaching for them to improve? In many instances, I believe the answer is "no" and that this is a disservice to the people we lead. We might limit feedback to our formal process for meeting once (or possibly twice) a year in a frequently awkward manner. Often, we avoid the difficult conversations that are necessary for providing useful feedback.

Building Women Leaders: A Blueprint for Women Thriving in Construction,
First Edition. Gretchen Gagel.
© 2025 John Wiley & Sons, Inc. Published 2025 by John Wiley & Sons, Inc.

I believe that the skill of effectively providing feedback and coaching is one of the greatest gifts a leader can provide to those they lead and coach. Most people appreciate feedback if it is given in a constructive way that conveys the care you feel for that person and your responsibility as a leader for their success.

> *"I love when someone gives me feedback, respectfully, as I can do something with that."*
>
> (Neuscheler, 2023)

As you read this chapter, I encourage you to thoughtfully reflect upon your leadership skills related to clarifying your expectations with those you lead and influence and providing them with feedback. How adept are you at coaching the members of your team, and even your peers? In what ways might you improve, given the ideas shared in this chapter?

Clarifying Expectations and Providing Direction

Years ago, I was coaching a CEO, and this person was frustrated because they had asked their executive team to improve performance in several key areas—and in their opinion, that was not happening. I asked the CEO, "Do these people clearly understand the timeframe of your expectations? Do they understand how fast you would like that needle to move?" Upon reflection, their answer was, "Probably not." People have less chance of meeting our expectations if we are not transparent and clear in communicating what a good job or good action looks like. Earlier, we discussed the acronym SAM—to "set direction, align, and motivate" (Kotter, 2012). Setting direction, clearly identifying what the finish line looks like, and defining the metrics of success is important for each of the people you have a relationship with as a leader.

I believe expectations are defined as "the desired results we expect from another, the level of performance and action we anticipate experiencing, and include the behaviors to get there." The quote I shared earlier by Burns mentions "mutually agreed upon goals" (Burns, 1978). I find that leaders at times unknowingly carry their expectations around in their heads without explicitly sharing these expectations with others. We are human and we make assumptions. "Of course, they know what I want," or "Of course, they know they are letting me down." Does this sound familiar?

> Clarifying expectations means:
>
> - **Verbally articulating expectations.** This includes spelling out details on desired actions, outcomes, timeframes, etc. to those you interact with, including specific measurable objectives.

- **Documenting these conversations in writing.** This could be a follow-up email. "Just following up with an email to confirm that we've agreed you'll have XYZ project schedule to me by ABC date."
- **Communicating changes in expectations.** "I know I said I need that project update by Wednesday, but the client now wants it on Monday. Is that doable, and how can I help?"

Clarifying expectations also involves enlightening people as to the context of their work so that they are an informed team member and understand the "why" behind your expectations. For example, I ask that my University of Denver Online MBA students answer their online asynchronous questions 24 hours prior to our live virtual session together. Why? Because if all 15 of them answer 30 questions just before class, I have no chance of reading each answer and modifying the agenda for our virtual discussion to meet their unique needs. That is the "why" of my expectation.

Your ability as a leader to clearly and concisely articulate your expectations to those you lead and influence ensures that you are aligned on the important elements of success in your relationship. Taking the time to do so is worth your investment.

Activity:

- Think of those who work for and with you. Does everyone understand their roles and responsibilities?
- Do these people clearly understand your expectations?
- What are some of the great questions you might ask to help you understand if this is the case? Questions such as, "Do you understand your priorities for this week?" and "Tell me what a good job on this looks like for you."

Providing Feedback

Feedback (an important component of any relationship or system) is defined as "the transmission of evaluative or corrective information about an action, event, or process to the original or controlling source" ('Feedback,' 2024). Consider the example of adjusting the temperature of your morning shower (Meadows, 2016). You turn on the water, then test the temperature with your hand to provide "feedback" within the system as to whether to turn the water temperature control to the right or the left. You change the dial, and you test the system again. Work relationships are similar to this system. None of us are perfect. We have things we do well in our job, and things we may not do well. We have skills that come naturally, and we have skills that do not come naturally. Feedback provides us with an opportunity to adjust and improve our performance and then celebrate the successes we achieve because of the feedback.

Amy Sandler, colleague of author of *Radical Candor: Be a Kick-Ass Boss Without Losing Your Humanity* Kim Scott, shared these thoughts on the Radical Candor® framework of feedback: "*We can care personally about the people we work with as real human beings at the same time we are willing to challenge them directly. We tend to see things as one way or the other, I must either care or challenge. The invitation is that you can do both. The reason it's radical is that it is rare that we are both caring and challenging. One benefit of the model of 'Care Personally and Challenge Directly' is to think, do I need to move up on 'Care Personally' or do I need to move out on 'Challenge Directly'? Kim was working at Google and had just given this huge presentation. The leaders at Google were excited to hear how well her team was doing. After the meeting, Kim's boss said, 'You said "um" a lot. Did you know that?' and offered a speech coach for Kim. Kim would be the first one to say she's pretty stubborn. She said it wasn't a problem and she was too busy for a speech coach. To challenge Kim more directly, her boss said, 'When you say "um" every third word, you lose credibility.' Kim went to the speech coach and realized, 'Oh my gosh, I have been saying "um" every third word for decades! I've been giving presentations my whole career and nobody has ever told me this.'*"

(Amy Sandler, speaking to Gagel, 2024 Greatness Podcast)

I believe Kim Scott's framework makes a great deal of sense. When you give feedback to someone, you are giving them a gift, and this must come from a place of caring. We all make mistakes we are not aware of. We all walk around with our sweaters on inside-out or broccoli in our teeth. It is our responsibility as a leader, a peer, and a teammate to have the courage to provide people with kind and effective feedback. If we have trusting relationships, they will appreciate it (Figure 15.1).

Providing feedback to people is an important gift, but may be challenging for a variety of reasons, such as:

- **Fear of an Emotional Reaction** – Sometimes, we may fear that someone will have an emotional reaction to our feedback because of the triggering caused by social pain (Hansen, 2018). If someone tells us we could be more effective by listening to all ideas before sharing our own, that might "trigger" the amygdala in our brain. If we do not have high emotional intelligence, if we do not have a "pause button," we might lash out at the person providing the feedback.

Figure 15.1 *Radical Candor: Be a Kick-Ass Boss Without Losing Your Humanity*, Kim Scott. The Radical Candor Framework is a trademark of Radical Candor, LLC.

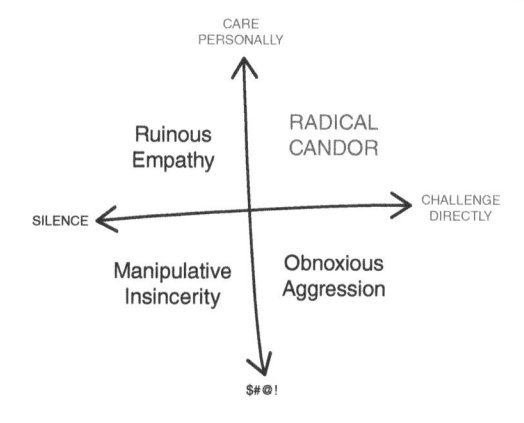

Experiencing emotional reactions to feedback might make us more hesitant to provide feedback to people in the future.

- **Lack of Time to Prepare and Deliver Feedback** – We are all so darn busy! How are we expected to carve out time to give meaningful, thoughtful feedback to anyone?! Especially if we are on a high-pressure construction project with schedule milestones to meet and activities to coordinate. There just is not time. I would argue that the time spent providing feedback and coaching is high return time as a leader—and intellectually, you probably agree. The challenge is in saying "no" to something else that creates the time for coaching.

- **Large Span of Control** – I spoke with someone at a global engineering firm who had 32 engineers reporting to them. How in the world are you supposed to effectively provide feedback to 32 engineers? I believe this is a rampant issue in the construction industry. We do not have enough effective managers and leaders, so we just keep piling people onto the teams of those who do it well.

- **Fear of Repeating a Mistake** – Most leaders have had at least one experience providing feedback in an ineffective way that resulted in backlash or a breakdown in their relationship with an employee. They might not have taken time to prepare, or maybe they did not have the facts exactly right, and they started providing the feedback anyway with disastrous results. I have made these mistakes. We all fall off the horse. The point is to get back on.

- **Fear of Getting It Wrong** – Do I have all the facts? Am I missing information? What if the feedback I'm about to give is not correct? These are the worries, the voices in our head, which may cause us to hesitate in providing feedback and coaching. We might worry that we are going to make a mistake with the facts of the situation.

Yes, giving feedback can be hard. Many things we do in life are hard but necessary. Be the leader, the peer, the teammate, who is effective in providing feedback to others. It will be appreciated. Here are some of my tips for effectively providing feedback:

1) **Keep to a Regular Feedback Schedule**

   I provide feedback to everyone who works for me on a quarterly basis (at a minimum). These meetings are scheduled out at least a year into the future, and these meetings are rarely (if ever) moved because doing so sends your team member a signal that they are less important than the reason you moved the meeting for. Sometimes, I will meet even more frequently. For example, when I am training a new assistant, I meet with them weekly for the first few weeks, and then monthly thereafter—in part to provide feedback. Keeping track of my crazy schedule can be a nightmare! How will they know if they are doing something wrong if I don't tell them? When you provide feedback frequently, it becomes an ongoing coaching conversation instead of an event or "oh my gosh, we're having a performance review" session. Feedback becomes a continuous, documented conversation regarding that person's strengths, opportunities, actions to improve upon these opportunities, and aspirations. I say "aspirations" because I always want to know where people want to be in five years, what skills they need to get there, and how I can help. I want each person to know I care about them and their life journey (Schein and Schein, 2023).

2) **Be Timely with Feedback**

   Many years ago, I was working with a construction project team in Idaho, and I would regularly check in with the owner's representative and contractor project manager by phone from Denver to see how things were going. For three months, I heard, "We're good," and then I started noticing tension in their voices. Next thing I knew, I was on a plane and locking them both in a room until they had dealt with issues they had been ignoring and the feedback they had not been giving one another for weeks. Growing up on a ranch, I think of this as getting too many "burrs under your saddle." They worked it out, but it was a much more difficult conversation, given how long they had waited to provide feedback.

   If something happens and you feel you need to provide feedback about that action or behavior to that person, I recommend you not procrastinate in doing so. Mickey Connelly calls these "point easy" conversations versus the "point difficult" or even "point crisis" conversation that occur when feedback is delayed (Connolly and Rianoshek, 2002). It is usually much easier to have the conversation right after something happens. If you wait, it is either an awkward conversation or perhaps the mistake has been repeated to the point that you are "triggered" with no pause button and lose it! (Figure 15.2).

   Acting too soon after the action or behavior may be a mistake as well. If something happens that really triggers you as a leader, you may need time to hit the

Figure 15.2 The Point Easy Conversation; Mickey Connelly, Conversant.

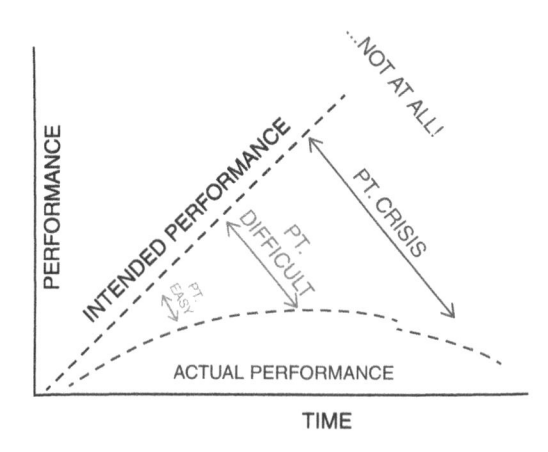

"pause button," or you will risk an amygdala hijack and knee-jerk reaction to the behavior (Hansen, 2018). For example, the project engineer who works for you enters the jobsite trailer and says she knows she was supposed to have the critical path schedule for Building 7 done today, but she forgot, and it is not done. You know you are triggered by people not keeping their commitments. You can "lose it" in an unemotionally intelligent way, or you can take a deep breath, thank her for letting you know, and ask her when it will be done. Later, you can sit down with her and have a well-prepared conversation to address this issue—especially if this is a reoccurring behavior or perhaps a capability challenge.

3) **Adequately Prepare**

I once spoke to a large group of energy engineers in Philadelphia, and the speaker before me was a wounded war veteran with a dog. Such a hard act to follow! I clearly remember one important point he made during his comments: as a sniper, if the bullet is off by a hair leaving the gun, it will miss the mark by several feet at the target. I feel the same way about performance feedback. If you start off on the wrong foot in the beginning of a performance feedback session, it is not going to go well. Most people do not spend nearly enough time preparing for performance feedback meetings. A one-hour performance feedback meeting requires a minimum of 30 minutes of preparation.

Here are the things I think about during my preparation:

- **What outcome am I seeking?** Is it a change in behavior? An agreement that we have an issue? Specific actions to improve? A celebration of a success?
- **What are the first two questions I need to ask?** What are all the questions I might possibly use during the feedback session? Notice I say "questions." This goes back to whether you are a telling leader or an asking leader (Schein

and Schein, 2023). I might ask things like, "How do you feel things have been going?" or "How did you feel about the report you did for client XYZ?" Often, employees will tell me what went wrong instead of me having to tell them.

- **Where should the meeting take place?** This seems simple, but having once started to provide someone feedback on a rental car bus in Detroit after a missed client deadline, I can say this is not always thought through! Recently, I was providing feedback to a CEO I am coaching, and I specifically asked that we be in a room with a large whiteboard where employees could not see us or the board, so that I could freely write on the board without others seeing the feedback. Location is important.
- **Am I in the right head space to provide this feedback?** I think about whether I am "triggered" or calm. Am I genuinely focused on improving the performance, or am I still angry about something? I need to be focused, prepared, and emotionally present to do a good job of providing feedback.

Using the same preparation and script does not work for every employee. The CEO of a nonprofit and I were having separate meetings with two Board members—both of whom we were asking to step down after their first term on the Board, as they were not a good match for the skills we needed. The first meeting went well. The second did not because the CEO used the same script, and a different conversation was necessary. It was a learning lesson for us both.

4) **Demonstrate Care**

I once asked Edgar Schein if you could teach leaders to care (a question I have deeply pondered). I eagerly anticipated his response, which was, "What do you think?" No, I want to know what you think! I believe that people do not follow leaders who do not care about them, and I am not certain you can teach leaders to care. As a leader, you must consider these questions: Do you care about this employee? Are you genuinely interested in helping them live the best version of themselves at work? How will you demonstrate this during the performance feedback session?

"*I think caring can be learned and I think we all have caring inside. It is through inquiry not through advocacy. You can't come in and hit someone over the head to say you need to care about people, or they won't follow you, even though it's true that they need to care. You have to ask them, 'What do you care about?' and if you care about this mission, this goal, this project, this job, it's going to take the willing participation of others to make that happen, and that starts with caring.*"

(Amy Edmondson, speaking to Gagel, 2019 Greatness Podcast)

I show employees that I care as a leader in part by regularly telling employees that they are a human first and an employee second, that my first concern is their happiness and wellbeing at work, and I mean it. As a leader, it is important for you to establish that you care about that person, their happiness, and their success.

> As Badar Alam, Corporate Capital Project Construction Competency Leader at The Chemours Company, shares, "*I worked for a woman in our industry that generated a level of respect, a tremendous leader and I felt a level of meaningful care and concern. I was not just an employee. She knew our families, the challenges we were facing. She wanted to connect with you first personally, 'How is your wife?', 'How are your kids doing?'*"
>
> (Alam, 2023)

5) **Focus on People AND Process**

Sometimes when there is a breakdown in a person's performance, it is actually a process breakdown. As Edward Deming said in 1993, "A bad system will beat a good person every time" (Hunter, 2015). For example, I was meeting with one of my executive assistants. We had recently sent an invoice to a client without including my travel expenses. I explained to her the importance of ensuring that we included travel expenses on all invoices, and that if these were not included, these expenses essentially came out of my personal pocket as the owner of the company. This should sound familiar as the first phase of developing psychological safety—framing the work and what is at stake (Edmondson, 2018). My next questions for her were, "What's not working in the process of me sharing my travel expenses with you?" and "What should we do differently in the process?" This is the second part of the psychological safety framework, inviting participation to solve the problem and demonstrating situational humility, as in, "I don't have all the answers, let's solve this together" (Edmondson, 2018). Most people want to do a good job. Is there something in the process that is broken and causing mistakes?

I've found swim lane diagrams to be quite effective in documenting process and diagnosing process breakdowns. Swim lane diagrams are process diagrams that include horizontal lanes representing each person or group of people in a process (MindTools, n.d.). Our team was working with a group of people responsible for delivering a significant capital program that was facing challenges in meeting their schedule and budget. After creating a swim lane diagram for the construction project delivery process, we realized there were more than a dozen people responsible for the project's success and no one point of accountability. The exercise of creating the swim lane diagram helped us effectively diagnose where process inefficiencies were creating personal performance challenges.

6) **Be Specific**

Effective feedback is specific because vague feedback is rarely useful in that you are not helping the person understand what specific skill or behavior is causing an issue. This is where sticking to the facts and avoiding assumptions is critical, and please avoid pulling other people's opinions into the conversation. It is never helpful to say things like, "Everyone thinks you are doing a poor job of focusing on the details in your project reports."

The Center for Creative Leadership has a performance model that I like called SBI—situation, behavior, impact (Gentry and Young, 2022). To use the model, you describe the situation (in yesterday's meeting); the behavior ("I felt you talked over Kate when she was trying to share an idea"); and the impact ("It caused Kate and others to shut down and not share their ideas"). While I like this model, I also suggest using questions at times instead of "telling" (Schein and Schein, 2023). For example, you might ask the person how they felt the meeting went. If they did not notice that they were talking over Kate's ideas, you might point that out and then ask, "What do you think the impact of that was on Kate and others in the meeting?" I also feel it is effective to ask people what they might do differently in the future, and if I have their permission to provide additional feedback in the future as a feedback loop.

7) **Actively Problem Solve and Create an Action Plan**

Once you have agreement on the issue at hand (a skill or behavior that needs to improve or change), demonstrate that you want to work **with** this person to solve the problem. You might demonstrate this by saying things like, "Let's work on this together," or asking, "What steps might we take, together, to solve this?" You might brainstorm ideas together, select an option, and determine the specific steps each of you will take to solve the problem. This could involve implementing new steps in a process or ensuring that the new behavior is understood. Be certain to leave the meeting with a documented action plan for steps both of you will take to help the situation. I always try to ensure that as a leader, I have steps as well, as a performance issue is rarely a one-sided affair. I keep a running document of meeting notes for each person who reports to me. This makes it easy for me to review the notes and actions from the last meeting and prepare for the next one.

8) **Anticipate an Emotional Response**

I have experienced situations where I am providing feedback and someone becomes emotional, or perhaps even begins to cry. The first time this happened, I was really taken aback. I was not saying anything horrible, just pointing out something that needed to be done differently. What I later realized was that this person was overwhelmed by work and was not talking to me about these feelings of being overwhelmed. I was not displaying emotional intelligence in understanding her emotions and how she was feeling, and I was completely oblivious to her stress. Whether that emotion is disappointment in letting you down, anger, or defensiveness, taking some time to anticipate and prepare for that emotional response is helpful.

Providing feedback to those you lead and influence is a gift and a skill that must be cultivated over time. Rarely are we perfect at providing feedback. Celebrate the times where providing feedback goes well and learn from the times that are perhaps more challenging. People appreciate feedback that is given with care and a genuine concern for their success.

Activity:

- How would you rate your ability to provide feedback to others? Do you need to be more "caring," or more "courageous"?
- Which of the tips for giving great feedback do you do well? What one tip might be an area for improvement?
- What one step could you take to more be more effective in providing feedback to others?

Conclusion

During one of my bi-annual anonymous reviews by those who worked for me, it was pointed out by more than one person that I had done an inadequate job of clarifying my expectations when assigning tasks and projects. This was true and due to a combination of being busy and having very little "C" in the DiSC profile—that attention to detail. I would fling a brief email or fire off a passing comment about something that needed to be done for the company or for a client, and guess what? I often did not receive what I was asking for. This caused a tremendous amount of frustration for both me and the person who had done what they thought was a good job. I took on board the feedback, developed an improvement action plan, and reviewed it with my team for feedback; then, I developed an ongoing system of feedback on this item to see if I was improving. It worked, and I improved.

Imagine what the world would look like—your world as a leader and the people you coach—if everyone knew exactly what the expectations were for their job, what skills you expected them to have (such as effective communication, listening, empathy, or scheduling), and what their "score" was for each skill. Imagine if we were adept at celebrating their strengths and developing strong action plans to work on their areas for improvement. Having mature, calm, and regular coaching sessions with your team is a key responsibility as a leader. You will not be perfect at providing feedback. Celebrate when it goes well, be kind to yourself when it does not, and keep working at it.

References

Alam, B. (2023). Interview by Gretchen Gagel [Zoom]. 25 October.

Burns, J. M. (1978). *Leadership*. New York: Harper & Row Publishers.

Connolly, M. and Rianoshek, R. (2002). *The communication catalyst*. Fort Lauderdale: Kaplan Publishing.

Edmondson, A. C. (2018). *The fearless organization: Creating psychological safety in the workplace for learning, innovation, and growth*. Hoboken, NJ: John Wiley & Sons.

'Feedback.' (2024). *Merriam-Webster Online Dictionary*. Available at: https://www.merriam-webster.com/dictionary/feedback [Accessed: 12 June 2024].

Gagel, G. and Edmondson, A. (2019). *Amy Edmondson discusses her book The Fearless Organization on psychological safety* [Podcast]. 30 October. Available at: https://open.spotify.com/episode/6joytIVhV3XlGFZx6rcodC [Accessed: 28 April 2024].

Gagel, G. and Sandler, A. (2024). *Amy Sandler discusses* Radical Candor® *and giving feedback* [Podcast]. 14 June. Available at: https://open.spotify.com/episode/0mh27h2fmCrBoYTSz4iFH2 [Accessed: 14 June 2024].

Gentry, B. and Young, S. (2022). 'Use Situation-Behavior-Impact (SBI)TM to understand intent,' *Center for Creative Leadership,* 18 November. Available at: https://www.ccl.org/articles/leading-effectively-articles/closing-the-gap-between-intent-vs-impact-sbii/ [Accessed: 12 June 2024].

Hansen, K. (2018). *Traction: The neuroscience of leadership and performance*. Sydney: EnHansen Performance.

Hunter, J. (2015). 'A bad system will beat a good person every time,' *The Deming Institute,* 26 February. Available at: https://deming.org/a-bad-system-will-beat-a-good-person-every-time/ [Accessed: 12 June 2024].

Kotter, J. P. (2012). *Leading change*. Boston: Harvard Business School Press.

Meadows, D. (2016). In a world of systems. 4 March. Available at: https://www.youtube.com/watch?v=A_BtS008J0k [Accessed: 12 June 2024].

MindTools. (n.d.). *Swim lane/Rummler-Brache diagrams*. Available at: https://www.mindtools.com/abxyani/swim-lanerummler-brache-diagrams [Accessed: 12 June 2024].

Neuscheler, K. (2023). Interview by Gretchen Gagel [Zoom]. 13 December.

Schein, E. H. and Schein, P. A. (2023). *Humble leadership: The power of relationships, openness, and trust*. 2nd edn. Oakland: Berrett-Koehler Publishers.

Part IV

"We" – Building Strong Relationships – Conclusion

At the time I joined the company, every Ralson Purina trainee (regardless of whether their eventual job would be in accounting, logistics, or operations) was required to work in the plant for the first four-to-six months to learn every job, from the unloading of ingredients to the shipment of pet food out the door. You were not just an observer; where possible, you were doing the job—be it swinging a sledgehammer at a rail car of flour or emptying 50-pound bags of Puppy Chow coating for eight hours. You also worked every shift, as we operated 24 hours a day. At the end of this training period, each trainee turned in a report of the operations, and if you had proven your ability to work with the people in the plant, you were then promoted to a permanent position. About 40% of the trainees were let go because the organization recognized that the critical skill for all managers, no matter their position, was the ability to build relationships with people.

I encourage you to continue reflecting upon your relationship-building skills long after you have read this book. If a relationship is not going exactly as you planned, take the time to think about the "why" and what steps you might take to strengthen the relationship. Is it a lack of trust? Are you missing an emotion they are experiencing? Have you wielded power in an inappropriate way? I also encourage you to be kind to yourself and realize that every relationship is a two-way street. You will work to build relationships with people who are incapable of holding up their end of the deal, such as narcissists, emotionally unstable people, and even people who are not worthy of your trust. Give it your best and know when to step away from these relationships. I see a great leader in all of you!

Building Women Leaders: A Blueprint for Women Thriving in Construction, First Edition. Gretchen Gagel.

Part V

"Us" – Leading and Influencing Teams and Organizations – Introduction

When you transition to the level of leading a team, "team of teams," department, business unit, or an entire organization, the elements of leading individuals we have discussed continue to be critical to your leadership success. In addition, the importance of several additional elements of leadership (including applying previously discussed elements such as purpose to this broader perspective) rises to the forefront. As we have discussed, leadership is fundamentally about creating movement and change. While enabling individual change through strong relationships is a critical component of success as a leader, understanding teams and organizations as complex systems and knowing how to lead and effect change within these systems is important as well.

> As Harvard professor, author, and speaker Michael Beer shares, *"Underlying this whole thing is the notion that organizations are systems. And all systems have complexity in them. There are multiple facets. There is no way for any single individual to understand all the facets. To create a fundamental transformation, you need a deeper meaning. It must be connected to the emotional components of the organization, the cultural and emotional components, and if you do not get those connections, you are just going to get a bunch of hard changes and you will not get that trust and commitment"*
>
> (Michael Beer, speaking to Gagel, 2020 Greatness Podcast)

Based upon research and my observations of hundreds of construction teams, I have created my list of the elements that make teams (and in turn, great organizations) thrive.

Building Women Leaders: A Blueprint for Women Thriving in Construction,
First Edition. Gretchen Gagel.
© 2025 John Wiley & Sons, Inc. Published 2025 by John Wiley & Sons, Inc.

- **Defining Purpose, Goals, and Measurable Objectives** – Great teams and organizations are aligned on the vision, the purpose of the organization, and the "why" of existence. The goals are clear, as are the measurable objectives that signal our progress.
- **Building Organization Culture and Values** – Great teams and organizations take time to understand the organization's culture and the values that culture is grounded upon; and ensure that leader behaviors align with these values and support the organization's culture.
- **Executing Strategies and Tactics** – Great teams and organizations have a strong plan of action, including high-level strategics that drive achievement of the organization's purpose and goals, along with operational tactics that make the strategies actionable.
- **Utilizing Teams, Norms, and Effective Meetings** – Great teams and organizations understand how to form and enable high-performing teams that are adept at establishing and adhering to norms and social contracts, as well as how to and train these teams in effective behaviors such as eliminating the bad meetings that are killing our high-performing cultures!
- **Attracting, Hiring, and Retaining Diverse Talent** – Great teams and organizations attract and select diverse talent with the right skills and values, and effectively engage this talent in the work such that they thrive and stay.
- **Creating Agile Teams and Organization** – Great teams and organizations understand the need to build change management as a core competency to remain competitive, and are adept at developing agile organizations that morph in reaction to the ever-changing business environment.

Entire books have been written on each of these topics. My intent here is to discuss each topic within the context of women leading in the construction industry and to inspire you to dig even more deeply into each subject.

Reference

Gagel, G. and Beer, M. (2020). *Michael Beer discusses his book Fit to Compete on the need for honest conversations* [Podcast]. 22 June. Available at: https://open.spotify.com/episode/1qf81DiFGkMMf4oEs1Lznw [Accessed: 28 April 2024].

16

Defining Purpose, Goals, and Objectives

Introduction

When Alice Hoffman and I worked together on the Baltimore Ravens Stadium project many years ago, we started building the construction project team by defining the purpose, goals, and objectives of the project. We aligned the entire executive team, ensuring that everyone agreed on each element, and continued to focus on the purpose, goals, and objectives throughout the project. The results were outstanding, and the team even published a CD of songs created during each of their monthly report-outs.

> *"Most of us are just waiting to be a part of something larger than ourselves. We want to be part of a team that is doing something great. If you give someone that opportunity and create the conditions, psychological safety and some process discipline, people will surprise you, they will show up and bring more than you ever thought they could bring to that opportunity."*
>
> (Amy Edmondson, speaking to Gagel, 2019 Greatness Podcast)

In his seminal book *Good to Great,* Jim Collins shares the results of his team's studies of various companies to determine what differentiates "good" companies from "great" ones (Collins, 2001):

- Great organizations understand why they exist and have an articulated vision of the difference they want to make in the world.
- Great organizations understand the goals they hope to achieve.
- Great organizations develop measurable objectives to help themselves understand whether progress is being made.

The importance of thinking about purpose, goals, and objectives does not begin and end at the top of the organization. Each business unit, department, and team must also think about their purpose, goals, and objectives to ensure that the strategies and tactics and actions of each member of these groups are aligned. I discuss two additional critical components of great teams and organizations (values and organization culture) in the next chapter.

As you read through this chapter, consider your thoughts on the purpose, goals, and objectives of your organization and how well these are understood by your team. Consider your team's purpose, goals, and objectives, as well as people's understanding and alignment around these elements. Reflect upon how you might strengthen people's commitment to them.

Team/Organization Purpose

For decades, I have remembered DPR Construction's purpose, which was developed in 1992 with the help of Jim Collins: "We Exist to Build Great Things"® (DPR Construction, n.d.). This purpose is DPR's north star—the statement that guides all they do as a company that executes billions of dollars of construction. The intent may change, the strategies and tactics may change, and the markets DPR Construction serves may change, but this purpose will endure. Great organizations have an enduring purpose that defines what Chris Worley calls the "identity" of an organization—the reason we exist and the change we want to make in the world (Worley, Williams, and Lawler, 2014).

> "All organizations start with 'why', but only the great ones keep their 'why' clear year after year. Those who forget 'why' they were founded show up to the race every day to outdo someone else instead of to outdo themselves. The pursuit, for those who lose sight of 'why' they are running the race, is for the medal or to beat someone else."
>
> (Sinek, 2009)

It's not enough to define an organization's purpose via a purpose or vision statement. This purpose must come alive via leaders' actions and be understood such that business units, departments, teams, and individual employees can align their purposes as well. If DPR exists to build great things, how does that translate to their project execution people, their estimating group, and their marketing department? How can each team of people create strategies and tactics that align with this purpose?

I became CEO/President of Continuum Advisory Group soon after the company was formed, and we put a great deal of effort into defining our purpose, "To transform the worldwide building and construction industry through revolutionary innovation." I remember there being a debate about the word "revolutionary," and there were hard conversations to drive us to consensus. This purpose guided our choices and allocation of resources. We wanted to ensure that our purpose was present daily, so we printed it on the back of our business cards and included it as a tag line on our letterhead. I had a slide made of the purpose statement, and often included it in public and client presentations when discussing our company. I also spoke about our purpose frequently with our team. To me, an organization's purpose is only important if it is talked about and used as a guiding beacon for the choices an organization makes.

This becomes even more critical in what I describe as a world of "rapidly reconfigurable resources," where teams are constantly morphing. This constant shifting of resources and reforming of teams are easier if we are aligned with our purpose as our guiding compass.

> *"I realized there were more and more times where I saw people were working with other people but not in stable intact teams. The boundaries of their teams were fluid and porous, and from a conceptual point of view I realized I might be onto something important in that more and more of the work is being done through teaming rather than in and by stable teams. That becomes a capability that everyone needs to develop more of, to get up to speed quickly with someone, to understand, what are we trying to get done. What do you bring? What are you worried about? What are you up against?"*
>
> (Amy Edmondson, speaking to Gagel, 2019 Greatness Podcast)

Often, an organization's purpose is codified in a vision statement articulating how that organization will make a difference in the world. For example, Google has the vision "…to provide access to the world's information in one click" (Dash, 2023). How do vision statements and mission statements differ? If the people understand why your organization exists—the difference you are trying to make in the world—then you have alignment on your purpose, the finish line for the boat, the vision. Mission statements more frequently capture the "how" of achieving the vision.

As either an informal or formal leader, you have the power to ensure that: (1) the purpose or vision of your organization is visible and frequently articulated to those you interact with; (2) people grasp how the "why" of their department or team aligns with this higher organizational purpose; and (3) your actions and the actions of those you influence align with this higher purpose. Alignment on our purpose, our identity, and the vision of why we exist in the world is critical.

Activity:

- What is the purpose of your organization? Do the people you work with understand and live that vision?
- How does the purpose of your team align with the purpose of your organization? How can you, as a leader, bring that vision to life?
- Think about how you might create stories that help share the purpose of the organization and how everyone contributes to that purpose.

Team/Organization Goals and Objectives

My team was hired to work with one of the largest corporations in the US to help them deliver a multi-billion-dollar construction project as quickly as possible. Speed was of the essence because of demand for the product. This client reached out to their most trusted construction management (CM) and engineering partners and sole-sourced their services without a bidding process. Our team was brought in to help develop a three-year project culture focused on Lean construction practices and collaboration to ensure that deadlines and budgets were met while also achieving safety and quality goals.

My colleague and I started by meeting with the seven executives of the client, construction management, and engineering companies to reach agreement and alignment on the culture of the project (values, norms, and behaviors) and the goals and measurable objectives of the project. One early conversation on our goals and objectives stands out. I asked the group of seven executives to tell me the key goals of the project, and one item we discussed was "substantial completion by XX date." One at a time, I asked each of the seven project executives what "substantial completion" meant, and I received seven slightly different answers. It took some time for us to agree on and document what "substantial completion" meant, along with the corresponding date. This discussion was critical to ensuring that we were aligned with the goals and measurable objectives of the program and could clearly convey these goals to the balance of the project team.

A goal is defined as "**1**: the end toward which effort is directed" ('Goal,' 2024). What is the finish line? How do we know when to celebrate? A construction organization might set a goal to reach the ENR Top 100 list (top-line revenue). An organization might set an EBITDA (earnings before interest, taxes, depreciation, and amortization) goal or profit margin goal, or a safety goal to be one of the safest companies in the industry. Each business unit, department, team, and individual's goals should then aligned with these high-level goals.

A leader is responsible for ensuring that each team member clearly understands their "mutually agreed upon goals" (Burns, 1978). As a leader, you understand if alignment exists by talking about these goals frequently and documenting goals with a team dashboard that regularly and visibly shares the goals and measurable objectives of the team. "Alignment conversations" remove the nuances of understanding we all have about the world and what is important—what is at stake (Connolly and Rianoshek, 2002).

> I had the good fortune to hear Billy Bean, the person characterized in the movie *Moneyball*, speak at the Construction Industry Institute (CII) annual conference (Lewis, 2004). Billy was made famous by applying statistics to baseball for the Oakland A's, a professional baseball team that was experiencing a salary cap challenge because of too many highly-paid players on the roster. The goal was to improve the team's performance and remain below the salary cap. Billy and his associate, Paul DePodesta, created a system for analyzing the statistics of younger, lower-salary players to select players based on the skills needed by the team. People not only doubted the system, but Billy also spoke of receiving death threats for trading away high-performing employees. But their system worked, and they had an extremely successful 2004 season.

An objective is defined as "something toward which effort is directed: an aim, goal, or end of action" ('Objective,' 2024). Objectives are measurable and establish our progress to our goals. I believe in the saying, "What gets measured gets managed," which is at times attributed to Peter Drucker (although some debate this attribution). For example, in the construction industry, we spend a great deal of time defining our safety objectives. We set a goal as an industry to ensure that everyone goes home in the same condition they came to work. Agreeing as an industry on key safety objectives and metrics was a big step forward, as we could see consistent measurement across companies and measure progress.

Measuring performance at every level of the organization is vital to your ability to lead. This is not about measuring everything; it is about measuring the objectives that are important to the success of each team, each department, and each business unit in order for the organization as a whole to achieve sustained performance. We all know what happens when one thing goes wrong with our car, such as a tire going flat or the alternator breaking. The car cannot function because one piece of the system is broken. Teams and organizations are similar systems. All people, all teams, all departments, and all business units must be well-functioning as defined by critical performance metrics.

When we worked with an energy company to reduce their annual capital program underspend, the goal was to achieve the capital budget that the CEO communicated to Wall Street and thereby enhance investor confidence. The measurable objective was to miss the budgeted capital expenditures by less than 2% each year. To achieve that measurable objective, we created a swim-lane diagram of the entire process from the inception of a capital project through to the commissioning of a pipeline. Each lane in the diagram represented a person, team, or department that was responsible for the steps in the process. One breakdown in the process meant failure for the high-level objective. We created measurable objectives for each "lane" of the process. For example, the right-of-way department had a measurable objective of how many months prior to construction they secured the land. This demonstrates how higher-arching goals and objectives cascade down through different departments, teams, and individuals.

It is critical that objectives are measured and communicated such that everyone's actions are aligned to the common goals. I am a huge fan of company, business, team, and even individual dashboards, which are concise visual representations of how we are doing on our goals and measurable objectives. I consider an organization or team's dashboard like a dashboard of a car, where you must have the right "dials" monitored to effectively operate the vehicle. These dashboards ensure that everyone is aware of the goals and corresponding measurable objectives, as well as our performance regarding these goals and objectives. If we are overperforming, it is cause for celebration. If we are underperforming, the dashboard provides feedback quickly so that we can make early course corrections. In management consulting, we knew the dials were utilization, billable rate, and satisfied customers, and that if we were achieving these goals and corresponding measurable objectives, our purpose of transforming the industry was achievable. On construction projects, these "dials" are typically physical and mental safety, schedule, budget, and quality, and we set measurable objectives for each.

While at FMI, Lou Bainbridge, Bill Spragins, and I worked to evolve "project partnering" on construction teams to a deeper level of collaboration throughout all of the project trade partners and employees—a process we called "High-Performing Teams." One key aspect of this was the development of productivity measures for each of the major subcontractors, with graphs to track these productivity measures that were posted on a board at the entrance to the jobsite. I remember one high-tech project we were working with that developed a metric for caisson drilling that was tracked on the jobsite board, and the subcontractor fell behind early in the project. Peer pressure is a powerful motivator, and because the entire project team had bought into the high-level goal (an on-time project) and each measurable objective (including how many caissons needed to be drilled a day), the subcontractor team picked up the pace and achieved the project goal and objective.

Connecting each employee's actions to the high-level goals and objectives of the organization is just as important. As part of the solution for the energy company mentioned above, the engineering group create a training video on how all employee's jobs and actions drive share price and shared it with everyone at every level—project engineering, field inspectors, everybody. It was effective in part because they had found a humorous way to explain the concept, and because it truly helped everyone understand how the actions of themselves, their team, their department, and their business unit all drove the achievement of the organization's goals and objectives related to share price.

Your actions as a leader to define the goals and objectives of your team in alignment with those of the organization are critical to the success of your team. The effective use of visual reminders of team performance are one way to ensure that team members understand their current levels.

Activity:

- What are the high-level goals of your organization? How do the goals of your business unit/department/team mesh with these goals?
- Are your goals visible and understandable? What could you do, as a leader, to ensure that everyone is aligned on our goals?
- What are the measurable objectives of your team? Department? Organization? Is everyone clear on these?
- Are you effectively utilizing team, department, and organizational dashboards to communicate your goals and measurable objectives? How are you utilizing these tools to celebrate success and make course corrections on your strategies and tactics?

Conclusion

Our team was working with a global manufacturing company to help them implement Lean construction practices on every global construction project. One day, I was speaking with their head of engineering and construction, and he shared a frustration. Each plant and project team had been "hoarding" capital project money during the year, which was then returned unspent at the end of the year due to the perceived importance of not exceeding project budgets. The actual goal was to spend the entire budget and spend exactly the amount of capital necessary. The company had a goal to maximize shareholder return, as most listed companies do. Unfortunately, there was a major stock buy-back program that year, and the millions of dollars of capital that went unspent on projects could have been invested in that stock buy-back program, generating millions of dollars of return to the company.

This story illustrates how important it is to ensure that every person, every team, is aligned on why we exist as an organization and how we keep score. Cascading our purpose, the "why" of our existence, and our goals and measurable objectives through each level of the organization lays the foundation for individual, team, and organizational performance, including the achievement of goals and objectives.

References

Burns, J. M. (1978). *Leadership*. New York: Harper & Row Publishers.

Collins, J. (2001). *Good to great: Why some companies make the leap... and others don't*. New York: Harper Business.

Connolly, M. and Rianoshek, R. (2002). *The communication catalyst*. Fort Lauderdale: Kaplan Publishing.

Dash, H. (2023). *Crafting a clear and compelling vision and mission: Why everyone needs one* [LinkedIn]. 4 March. Available at: https://www.linkedin.com/pulse/crafting-clear-compelling-vision-mission-why-everyone-harish-dash/ [Accessed: 13 June 2024].

DPR Construction. (n.d.). *We exist to build great things*. Available at: https://www.dpr.com/ [Accessed: 13 June 2024].

Gagel, G. and Edmondson, A. (2019). *Amy Edmondson discusses her book The Fearless Organization on psychological safety* [Podcast]. 30 October. Available at: https://open.spotify.com/episode/6joytIVhV3XlGFZx6rcodC [Accessed: 28 April 2024].

'Goal,' (2024). *Merriam-Webster Online Dictionary*. Available at: https://www.merriam-webster.com/dictionary/goal [Accessed: 13 June 2024].

Lewis, M. (2004). *Moneyball: The art of winning an unfair game*. New York: W. W. Norton & Co.

'Objective,' (2024). *Merriam-Webster Online Dictionary*. Available at: https://www.merriam-webster.com/dictionary/objective [Accessed: 13 June 2024].

Sinek, S. (2009). *How great leaders inspire action*. September. Available at: https://www.ted.com/talks/simon_sinek_how_great_leaders_inspire_action?language=en [Accessed: 13 June 2024].

Worley, C. G., Williams, T., and Lawler, E. E. III, (2014). *The agility factor: Building adaptable organizations for superior performance*. San Francisco: Jossey-Bass.

17

Building Organization Culture and Values

Introduction

Several years ago, an investment bank recommended that our team provide market research to a private equity-owned construction industry organization that was struggling to achieve their growth objectives. During our first meeting, the executive team explained that their business development people lacked focus and were chasing a multitude of potential new markets. The organization wanted to understand how to profitably grow by focusing on the most profitable, high-growth markets. Although we were hired to provide market research, I began every meeting with two slides—one on organization culture and one on organization ambidexterity (balancing resources between current and future business models)—to reinforce two critical points: (1) we could flood them with information on which markets to pursue, why, and how, but until the organization changed a culture that reinforced business development chasing every "bright shiny object" (including what they valued and how they compensated their people), nothing was going to change; and (2) in order to succeed in a new market, adequate resources must be allocated to that effort. This message was heard, and with our help, the company was able to quadruple their revenue in less than a year in the new market we had identified.

What our teams and organizations value (along with the ensuing culture of our teams and organizations) is just as critical as our purpose, goals, and objectives, as this defines "how" we go about doing our business. Our organization culture and values also speak to our brand and the talent we attract. Rarely have I consulted an organization or coached a leader when understanding organization culture has not played a critical role in achieving success. Organization culture—including our values—is the fingerprint that exists on everything we do within an organization and shapes all behavior.

Building Women Leaders: A Blueprint for Women Thriving in Construction,
First Edition. Gretchen Gagel.
© 2025 John Wiley & Sons, Inc. Published 2025 by John Wiley & Sons, Inc.

Edgar Schein argued to me that you must always put another word in front of the word "culture," such as "organization," "industry," or "geographic," because culture is complicated and specific to these contexts (Schein and Schein, 2019). For example, the concept of occupational culture is important in our industry. For you engineers reading this book, what do you think of when I say "architect?" And vice versa? Or when I say, "marketing person," or "accountant?" Words spring to mind based on unconscious bias and the perceived cultures of these occupations. Geographic culture is also important, and living in Australia has taught me many nuanced differences in our geographic cultures. Even different teams and departments within organizations may have different subcultures. These facets of culture make thinking about the concept of culture quite complicated. You might be working with a construction project team made up of people from different countries and occupations, and you are probably also working with dozens of organizations—each with their own culture. It is complicated! My remarks in this book are focused on organization culture, but I encourage you to continue reflecting upon the many nuances of culture.

As you read this chapter, consider that your level of understanding of an organization's culture is recognized and influenced. Take time to think about how your organization's values and culture are shaped by your actions as a leader.

Defining Organization Culture

Culture is defined as "**1a:** the customary beliefs, social forms, and material traits of a racial, religious, or social group; *also*: the characteristic features of everyday existence (such as diversions or a way of life) shared by people in a place or time; **b:** the set of shared attitudes, values, goals, and practices that characterizes an institution or organization; **c:** the set of values, conventions, or social practices associated with a particular field, activity, or societal characteristic; **d:** the integrated pattern of human knowledge, belief, and behavior that depends upon the capacity for learning and transmitting knowledge to succeeding generations" ('Culture,' 2024). I think of organization culture as the "tribal smell" of an organization. It can be difficult to define an organization's culture as an outsider, but when you are in the organization, you feel the vibe. There are typically signs of a team or organization's culture throughout the organization. For example, if I walk onto a jobsite and the lay-down area is highly organized, that conveys a project team culture of discipline. If I walk into a jobsite trailer and there are comic sayings up on the wall, I see that this team values fun.

While many frameworks of organization culture exist, Schein's framework makes a great deal of sense to me and is defined by these three elements (Schein and Schein, 2019):

1) **The artifacts of the organization** – These are the physical manifestations within the organization that define the culture, such as the office configuration or the way people dress.
2) **The espoused values of the organization** – These are the values we place on our website, the things we say we believe in as an organization, and what we stand for.
3) **The underlying assumptions of the organization** – These are the unconscious beliefs that drive what people think and how people behave within an organization.

1) **Artifacts**

Back in the 1990s, our team advised two big box retailers on how to improve the delivery of their construction programs. Thinking back, this is probably one of my earliest experiences noticing the artifacts of an organization. The first time we visited the first organization's corporate headquarters, our team was given a map on how to navigate the maze of identical office spaces and avoid getting lost. It was a sterile environment, cubicle after cubicle, with a stark and unwelcoming foyer—and an equally unwelcoming receptionist! Our first visit

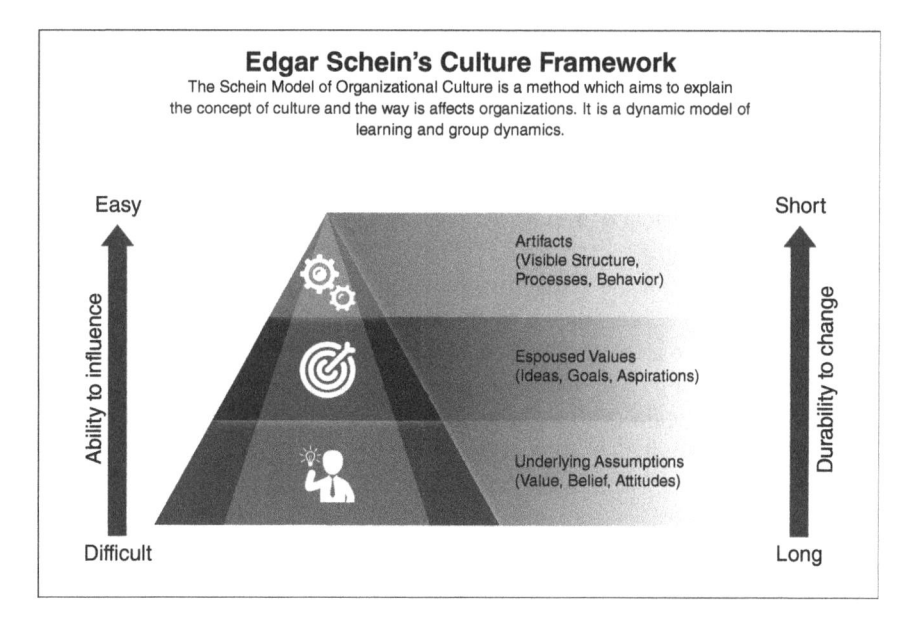

Figure 17.1 Edgar Schein's Model of Organization Culture. Source: Skyline Graphics/Adobe Stock Photo.

to the competing retail organization a year later was a stark contrast: a modern and welcoming reception area with a friendly receptionist, and warm, inviting office areas uniquely decorated with a fun and happy feel to the space. This is the essence of the artifacts of organization culture that encompass many physical aspects of an organization, including how people dress, how the offices are configured, and what appears on the walls (Figure 17.1).

Back when I first began reading *The Corporate Culture Survival Guide* by Edgar and Peter Schein, we were consulting an energy company that was regularly missing their capital budget spend by double digits. We tackled the internal processes that were causing project start delays and other, more tangible challenges, but we also knew there were some aspects of the organization's culture (including a lack of accountability) which needed to be addressed.

I suggested to the lead consultant that we try a new idea. We divided a group of 25 leaders from the engineering department into teams of five and asked each team to walk through their offices and take pictures of things they felt represented the organization. I do not believe I ever used the word "culture." When they returned 20 minutes later, we assembled the pictures taken by each team into a slideshow and asked the group to describe the pictures and why they had chosen that location. For example, someone had taken a photo of an engineer sitting in their office with headphones on, and used the words "lack of collaboration" and "isolation" to describe the picture. Every team had taken a picture of an old typewriter (some of you may have never seen one!) with a piece of paper taped above it that said, "Office Equipment Museum." They vehemently stated that it was still used for one form, and then spoke of the difficulty they had at times in letting go of the past. One group took a picture of awards on the wall and used words like "family" and "pride" to describe how this made them feel.

We created a list of the words the team used as they were describing each picture, and then asked, "Which of these words do we want to continue to use to describe the engineering department?" and, "Which words do we want to mitigate or change?" It was a very powerful exercise that helped the team understand their team culture without ever using the word "culture," instead relying on the "artifacts" of their culture to help them describe it.

When I think of the artifacts of culture in the construction industry, many examples come to mind. DPR Construction had one of the first open plan office spaces in the industry back in the 1990s, including no offices with doors and a phone booth you could go to if you needed to make a private call. I think of the many Kiewit highway projects I have driven by during an evening or weekend, with every piece of equipment lined up perfectly in order, and how this matched the disciplined culture of the organization. I also think of the mine site I visited where each of the manager's offices had closed blinds on the windows, and the one office without blinds assigned to me as a visitor had paper taped over the window. This conveyed a feeling of management isolation and secrecy.

Understanding how the physical artifacts of a culture reflect the values of an organization is an important insight to have as a leader. It requires being able to see the physical manifestations of an organization as items representing the values of the organization. Being present to the artifacts of a team or organization's culture is one way to gain insight into the culture.

Activity:

- Walk around your offices or jobsite with an eye to the artifacts of the organization. What do the artifacts tell you about the team or organization's culture?
- Brainstorm a list of the words that come to mind.
- When you think of these words, which ones are desirable that you want to keep? Which words are perhaps not desirable, and how would you change them?

---

2) Espoused Values

When I gather groups of leaders together for leadership development programs, I often ask them to recite the values of their organization. Typically, few people can. During one specific gathering of gas utility engineers, there were several people in attendance from the same company, and every one of them could not only recite the values, but also articulate what the words stood for and the actions and behaviors that supported these values in their everyday lives. I was blown away. This is a tremendous example of an organization that has embedded their values in every employee such that the values are lived daily.

Typically, the values of an organization stem from the values of the founder, and might evolve over time as leadership evolves. I think back to Doc Fails (the founder of FMI) and one of his iconic quotes, "Profit thrills, volume kills." Doc Fails started out by helping contractors become more profitable in a desire to positively impact the construction industry. He valued entrepreneurism, and that lived on in our company, with people like me starting new services and targeting new markets, all to improve the construction industry.

It is not enough to clearly define, articulate, and live by the values of your team or organization. One must also have zero tolerance for people within the organization who do not embrace these values. Too often, I have heard leaders say things like, "It's tough finding people, so we'll tolerate this person, he's the best we can find," or "Everyone knows that so-and-so is a bit rough around the edges, but we can tolerate his behavior." Keeping employees who do not live the values of the organization undermines your efforts as a leader to build an aligned culture, and you will lose good people as a result.

> Bill Boyar, Founding Shareholder of Houston law firm BoyarMiller, shares this story: "*We gathered our partners in the room several years ago because we were having a challenge with a partner who was harassing young associates. We made a commitment that if we are going to survive long-term, be a multi-generational firm, really be the place that we wanted it to be, we had to be willing to stand for our values. If you can't live consistent with the culture of this firm and behave consistent with our values, you cannot work here regardless of your position in the firm, regardless of your productivity, regardless of how much money you make, regardless of how big a practice you have, you cannot work here. We all said that and shortly thereafter we fired one of our most productive partners. That reverberated through our organization, and it was the sort of the turning point. As Jim Collins said, making the tough decision slowly is one of the most damaging things to an organization, and we learned to make those tough decisions quickly, we figure out early on, Gretchen, if somebody gets it. We give them a couple of chances to become accustomed to how we are, and if it's not going to work out, we cut our losses. At the end of the day, if you don't make those tough decisions, you're doing more damage for the long-term than you could possibly do by trying to rehabilitate somebody and failing.*"
>
> (Bill Boyar, speaking to Gagel, 2020 Greatness Podcast)

You may not be responsible for creating the values for your organization, but you are responsible as a leader for helping people remember, understand, and live these values. Leaders must talk about the values of the organization to create alignment and shape behavior. You must share and celebrate examples of behaviors that support the values, and there must be immediate and consistent ramifications for behaviors that do not support the values. Do you talk about the values and what each of them means? Do you celebrate examples of people living these values? Do you promptly address issues where people are NOT living the values? These are the important questions to ask yourself as a leader.

Activity:

- How many of your leaders of leaders, or team members, can remember the espoused values of your organization?

- When you ask them to articulate what each value means to them, are these ideas aligned with the organization's intent for that value?
- What ramifications (if any) are in place for people who behave in a way that is inconsistent with the values of your team or organization?

3) Underlying Assumptions and Shared Beliefs

During my PhD studies, we took a course that encompassed the study of organization culture. One Saturday, our professor asked us to go out in groups, select a restaurant for lunch, and then use Edgar Schein's framework of organization culture to determine the culture of that restaurant (Schein and Schein, 2019). I set off with two of my classmates and we chose a café at the mall in downtown Denver. We began by looking at the artifacts of the establishment and thinking of the words the décor suggested. When the waitperson arrived, after some initial banter, my fellow student turned to her and asked, "What are the shared beliefs of the organization?" She stared at him blankly, and I admit I had a chuckle.

I employed a different strategy and asked her, "So, when you are late to work, what happens?" Her answer was, "Not too much, you might have to clean the gum off of the underside of the tables." (YUCK!). I asked, "Does your boss do anything to make you feel appreciated?" Her answer was, "He's great. He takes us all to the baseball game once year." The waitperson's answers to these questions helped me understand the shared beliefs of the organization and its culture—I would use words like "relaxed" and "family oriented." This exchange became a bit of a running joke for our group, and my fellow student was good – natured about it, but it highlights that shared beliefs are difficult to uncover, and most employees have a difficult time answering a direct question about them.

We can plaster the walls and our website with the values we say we stand for, but do our actions support those values? Are the shared beliefs of the organization in line with those values? We can say we value work-life balance, but does my supervisor ask me why I'm leaving early? Or roll their eyes when I say I'm leaving at 5:00 to attend a kid's school play? Our actions and behaviors shape the organization's culture and reinforce our organization's values. As a leader, understanding the actual underlying assumptions and shared beliefs of the organization is critical.

> **Storytelling and Organization Folklore** Storytelling applies to how an organization tells its story to the world—our brand and how we market ourselves—and the internal stories we tell that become the "folklore" of our organization and define our culture and shared beliefs. This folklore does much to create the tacit knowledge and shared beliefs that are the underpinning of our organization's culture.

Here is an example of organization folklore that has stuck with me for decades. While I was at FMI, our investment banking group represented a company in a sell-side transaction with a contractual fee for the sale. After the sale was completed, the CEO of the client company approached Hugh Rice, the leader of the investment banking group, and said he did not think the fee was fair, even though it was the amount agreed-upon prior to the sale. Hugh then asked him what he did feel was fair, and then wrote a check for $140,000 to that client with a letter explaining that one of our most important values is customer satisfaction. Next, Hugh copied every employee on the check and the letter to send home a strong message about this philosophy, and that story sticks with me today.

"What I've found is a lot of people when they start thinking of brand storytelling, they think of one story and it's normally the founder's story or how the company was created. And I think that's important, the purpose of the company. Culture stories are stories that communicate your company values. This is normally leaders sharing personal stories of what the company values mean to them or the company sharing stories about employees living the values and demonstrating the values. I often write about the value of organization heritage over history. Many times, companies share their history, how we started here and it's very factual which can provide value, but they ignore the heritage, and the heritage is these folklore stories that convey our culture."

(Gabrielle Dolan, speaking to Gagel, 2021 Greatness Podcast)

Each of your actions as a leader, and each of the actions of those around you, creates the shared beliefs and culture of your team and your organization. Deliberate thought as to what these shared beliefs currently are and how you might want to influence them in the future is a worthwhile endeavor.

Activity:

- What do you consider to be the underlying assumptions and shared beliefs of your team? Do they align with the desired culture of the organization? If not, what might you do differently as a leader?
- Think of recent leader behaviors by you or others that reinforced the values of the organization.
- Now think of recent leader behaviors by you or others that might have sent a different message to employees. What could have been done differently?
- What are the stories—the folklore—of your organization? How do these stories reinforce the underlying assumptions and shared beliefs of the organization?

Measuring Organization Culture

How do you measure organization culture? Edgar would argue that it's not through a survey, although I do believe surveys can be helpful as one set of data points (Schein and Schein, 2019). Back in the 1990s when DPR Construction was expanding and opening new offices throughout the US, I was brought in to administer an annual employee survey of each office. DPR aspires to be one of the most admired companies in the world, and the culture of their organization is of critical importance to them. These surveys were one point of information that could point out if a new office was developing a culture that was not in alignment, or if an established office was going off course.

During my PhD studies and as a leader, I learned the value of what I call "boots on the ground leadership" versus "ivory tower leadership." The best way to understand the organization's culture is to be out amongst the employees. My daily walks on the floor of the plant at Beech-Nut Baby Food gave me incredible insight into our organization's culture. When I first arrived, we were experiencing a lost-time accident about every three days. I began to understand that our organization's culture accepted accidents as a way of life, and that we needed to start by stating our safety values and then acting upon the values by consistently pointing out unsafe behaviors and reinforcing safe behaviors. This shifted our organization's culture by creating new shared beliefs toward safety and significantly improved our safety statistics.

It is also important that leaders interact with external stakeholders to understand the culture of the organization. I recently read an article about a CEO who asked his entire senior leadership team to join him in answering their customer complaint hotline for a day and the incredible impact that had on their understanding of how they were perceived by their customers, as well as what that said about their organization's culture. Taking time to understand how your external stakeholders (shareholders, clients, suppliers) perceive and experience your organization's culture is important.

While it may be difficult to understand and capture an organization's culture in words, it is critical that a leader understands the organization's culture and how that culture is reinforcing the values and acceptable behaviors of the organization.

Activity:

- What steps does your organization (or you as a leader) currently take to understand your organization's culture?
- Is this effective? What more might be done?

Conclusion

A few years ago, I was excited to join a company. On paper, everything looked great. I admired the work they did, and I thought it was a great fit. I was wrong because of one critical factor: It was not a great organization culture fit for me. As I explained to the leader upon my departure, I am a trumpet player in a marching band, reliant on precise project management plans to run large engagements, and this organization was more of a jazz ensemble, making it up a bit as they went with less of a plan. Neither was right or wrong; it just was not a good cultural fit. My values did not align with the values of that organization.

I cannot overemphasize the importance of organization culture as you reflect upon great leadership. One organization's culture is not necessarily bad versus another being good. As a leader, you must define your organization's culture such that there is alignment in what is valued and how you want people to behave. If the people you hire are a great cultural fit, you will spend less time dealing with conflict that is rooted in different opinions about the organization's values and beliefs. How you utilize artifacts, espoused values, and shared beliefs to define and reinforce the culture of your teams, your departments, and your organization is critical.

References

'Culture.' (2024). *Merriam-Webster Online Dictionary*. Available at: https://www
.merriam-webster.com/dictionary/culture [Accessed: 13 June 2024].

Gagel, G. and Boyar, B. (2020). *Bill Boyar discusses the impact of creating a culture of authenticity* [Podcast]. 25 June. Available at: https://open.spotify.com/episode/
50hKp76xEGQDLbbKQ60482 [Accessed: 28 April 2024].

Gagel, G. and Dolan, G. (2021). *Gabrielle Dolan discusses her book Magnetic Stories* [Podcast]. 2 February. Available at: https://open.spotify.com/episode/
0uamofsvN18F1sWcjawMQm [Accessed: 28 April 2024].

Schein, E. H. and Schein, P. A. (2019). *The corporate culture survival guide*. 3rd edn. Hoboken, NJ: John Wiley & Sons.

18

Executing Strategies and Tactics

Introduction

Once your team or organization achieves alignment on purpose, goals, measurable objectives, organization culture, and values, determining the strategies and tactics to achieve successful outcomes is the next important step. Your ability to engage your team and peers in the development of effective strategies and tactics is critical to your success as a leader. The story of how Australian Olympian cyclist Anna Meares and her team developed a unique strategy to defeat her archrival, Great Britain's Victoria Pendleton, to win gold at the 2012 London Olympic Games is a compelling story of team strategy and tactics.

> *"Victoria Pendleton had been undefeated for six years internationally and no one had been able to work out how to beat her. The team that I worked with understood that if you continue to go about doing the same things without learning, tweaking, and adjusting, you ultimately fight for the same results. Our statisticians watched over three hundred hours of race footage of Victoria Pendleton to look for patterns of behavior. From that information we determined that when Victoria Pendleton had her rivals in front of her in a sprint match, when she could clearly see them and use the height of the track to her advantage, she won the majority of the races. However, when she was in front of her opponents and she had to look back at them, she won less often. We realized, if we're going to have the best chance of beating Victoria Pendleton, we needed to stop trying to race her at a best and most practiced position. In London we needed to ensure that Victoria Pendleton was in front of me in every sprint match.*
> *"At the world championships in 2011, I performed a skill known as the track stand which is literally coming to a stop on the bike if I was in the front position to force Victoria from the back into the front position, and I was successful. It was*

> *the first time in six years that Victoria Pendelton would not be champion. I did not use that strategy again for a year and a half until I met her in the final in London, and as history would show that strategy would become very famous and very successful for me to win my second Olympic gold medal."*
>
> (Anna Meares, speaking to Gagel, 2020 Greatness Podcast)

Anna and her team had come up with a strategy that allowed them to achieve their purpose and goal of winning an Olympic gold medal. Creating this new strategy involved analysis of past performance, brainstorming of new ideas, and above all else, risk taking. Anna could have looked foolish if it had not worked.

As you read this chapter, I encourage you to reflect upon how effective you and your team are in creating and executing strategies and tactics that drive performance. What ideas in this chapter might you adopt to improve this team capability? Are you and your team dedicating sufficient time and energy to the process of creating and evaluating strategies and tactics?

Defining Strategies

When I started the Owner Services Group at FMI, the vision was to help large corporations and government agencies with significant capital construction programs understand that low initial price did not always equate to low total delivered cost for a construction project or program. One year, a retail chain contacted me and explained that they had just remodeled 200 stores across the US by bidding every job and using about 50 different contractors. They were not certain that was the most cost-effective way to deliver that program (I agreed!) and asked our team to help them develop a new strategy for a similar remodel program for a different brand. We worked with their team to create regional alliances with contractors that eliminated the time and expense of a bidding process, along with a methodology for benchmarking and tracking the cost per square foot and other metrics (quality, schedule, and safety) with great success. Their goal of lowering the cost and reducing the time to remodel stores was achieved via a new contracting strategy.

A strategy is defined as "**a:** a careful plan or method: a clever stratagem; **b:** the art of devising or employing plans or stratagems toward a goal" ('Strategy,' 2024). I think of strategies as the "how" of us achieving our goals and objectives; the high-level plans a team, department, or organization use to achieve a goal and measurable objective. Setting strategy begins at the organization level with senior leadership utilizing extensive internal and external information, and typically some type of framework to develop an organization's strategies. Strategizing is not done in a vacuum, and strategies must be tested and revisited frequently. Given the increased volatility of the business environment, gone are the days of developing a strategic plan and putting it on cruise control for five-to-ten years. It is up to the business units and teams to then develop aligned strategies that support the high-level strategies of the organization.

> Involving your team and peers in the development of strategies is critical. People want to be in the know and feel that they are contributing to the team's success. While CEO/President of WFCO, I implemented "Team Time" every Thursday afternoon for two hours (I called it "Team Time" to make it sound more fun than "Strategic Planning Time!") Different from our tactical team meetings on Tuesdays, these strategic meetings engaged our team in focusing on our purpose, goals, and objectives through strategic conversations on a different topic each week. At first, it was difficult for the team to maintain a strategic, thirty-thousand-foot viewpoint on a topic (such as how we marketed ourselves to achieve our purpose or the timing of our next big endowment campaign). Over time, the team was trained to think strategically. In fact, three years into my career there, I hired a consultant to help us with a new strategic plan, and one of her first comments was, "Wow, your team is really good at thinking strategically." Our efforts had paid off.

I call the engagement of a team in developing strategies "baking the potato together," although in Australia I have changed this to "baking the meat pie together!" If you are a leader and you only take fully formed ideas to your team, there is less opportunity for them to buy into these ideas and you are not taking advantage of their diverse thinking. If you take an idea to your team that is 50–75% formed and ask them to "bake the potato" the rest of the way with you, the team will probably help you come up with a better idea (and this also leaves their fingerprints on the strategy, which achieves buy-in). After one particularly collaborative meeting with her construction team, one of my clients clapped her hands together and said, "I love baking the potato with my team!"

> Here are just a few of the strategic frameworks and models I have found useful when thinking about strategy at the organization, department, and team level:
>
> **SWOT Analysis**
> The tried-and-true SWOT Analysis dates to the work of Albert Humphrey in the 1960s and is a great starting point for periodically thinking **internally**

about the strengths and the weaknesses of the organization as it exists today, as well as looking **externally** to the threats to the current business models and opportunities for business model expansion (Teoli, Sanvictores, and An, 2023). The pros of this model are that it is simple and focuses both internally and externally. Conversely, the cons are that most times, it is a snapshot of where the organization is today, and it does not necessarily focus on the strengths needed for the future. It can also be challenging to encourage people to think about opportunities that are truly "out-of-the-box" ideas.

Jim Collins' Hedgehog Model

I am a fan of Jim Collins' book *Good to Great* and the "hedgehog concept" that drives an organization to success—three intertwined circles of: (1) what are you deeply passionate about? (vision, identity, purpose); (2) what can you be the best in the world at? (vision, identity, purpose, goal, strategy); and (3) what drives your economic engine? (goal, objectives, strategies) (Collins, 2001). I like this model because it causes an organization to drive toward excellence and being the best in the world at not just anything, but what they truly care about, and the framework connects this passion to the economic engine. The organization must monetize business concepts and operate profitably to achieve sustained success.

PESTLE Model

I learned about this model while teaching Strategic Management in the University of Denver's Online MBA program, and just recently used it again in a strategic planning process with a US-based engineering firm. This model also dates to the 1960s and the work of Francis Aquilar, and stands for the "Political, Economic, Social, Technological, Legal, and Environmental" aspects of the business environment (Reding, 2023). The PESTLE model is an effective tool for analyzing the external environment because it causes you to think about different categories of information and broadens people's thinking. The cons: it focuses on collecting the data, and people must realize that you also need to analyze the data to create information that is actionable—the "so what" of the data.

Galbraith Star Model

Yet another model from the 1960s (being a '64 model, I do think it was a great decade!), the Galbraith Star Model looks at the intertwined nature of Strategy, Structure, People, Processes, and Rewards (Sridharan, 2021). All too often, I experience an organization addressing one, two, or three of these elements, and a fatal flaw in one of the remaining elements sabotages a strategy. The organization may have wonderful people and be rewarding them for

> the right behaviors, but they may also have inefficient internal processes that cause frustration and turnover; or the organization may have the right people in the right structure, but a reward system that emphasizes the wrong behaviors. I appreciate this model because it touches on the organization as a complex system.

Each of these models contains elements that can assist you and your team in developing effective strategies that address an organization as a complex system and focus on achieving the desired goals and measurable objectives of the organization. Critical to the use of each of these models is the capability for leaders to have open conversations about the organization's abilities.

> *"One of the main underlying reasons for all strategic failures is the inability of the organization to have an honest conversation, to discover where it is not fit to compete, where it is not fit in either the values or the strategy. If you have those conversations, it inevitably reveals what needs to change."*
>
> (Michael Beer, speaking to Gagel, 2019 Greatness Podcast)

The ability to work with your team to develop the strategies necessary to enact the purpose, goals, and objectives of your organization is a critical leadership skill. It is not necessary for you to have all the answers. Leveraging the thinking of your team and having honest conversations about the organization's abilities and how these support the adoption of certain strategies are key.

> **Activity:**
>
> - What are the high-level strategies of your organization? Are the people on your team aware of these strategies?
> - How do the strategies of your team align with your organization's strategies? How might you, as a leader, ensure that this alignment exists?
> - How effective is your organization in engaging people in the development of strategies; in "baking the potato" together? What might you do as a leader to improve this capability?

Defining Tactics

Strategies must be operationalized into tactical business plans in order to be effectively executed. A tactic is defined as: "a device for accomplishing an end"

('Tactic,' 2024). While my intent is to maintain a high-level viewpoint in this book, it is important to mention tactics because this is how we actually get things done—through action! I'm reminded of the great book titled *Execution: The Discipline of Getting Things Done* by Ram Charan (Bossidy, Charan, and Burke, 2011). Operational plans detail the tactics each team, department, or business unit will execute to enact the strategies and achieve the goals and measurable objectives of the organization.

> Let us assume you are a project manager on a large construction project. One strategy to ensure that the project is completed on time is to make sure you, as the general contractor, maintain a good relationship with each of the subcontractors. How are you going to accomplish this? What tactics will you use to enact that strategy? You might meet with each subcontractor weekly or monthly to gain a sense for the strength of the relationship and how you help them achieve their goals. You might walk the jobsite with a different subcontractor superintendent each week. These are examples of the tactics you will use to achieve the strategy.

Tactics only work if the team is accountable for taking action. As underscored during our discussion of trust, reliability and capability are key components of taking action. It is not enough to write the plan down. Actions speak louder than the words on a page, and holding team members accountable for the effective execution of strategies and tactics is appropriate and necessary.

> When we worked with an energy company to improve their capital program delivery results, I asked the CEO for 90 minutes of time once a month for six months, and he agreed. We asked each of the five action teams working on performance improvement strategies and tactics to report-out monthly to the CEO on their past and future actions to help drive accountability. When I asked to see the team's presentations the afternoon before the first report-out to the CEO, I heard quite a few comments such as, *"He won't be there tomorrow, he's too busy,"* or *"Does he really expect us to be working on this with everything else we have to do?"* The next day, however, in walked the CEO, right on time to dropping jaws. These team leaders quickly realized they were going to be held accountable for making progress, and it worked—they did make progress, and quickly!

Organizations and teams serve both internal and external stakeholders with strategies and tactics that are aimed at achieving the company's objectives. Let's take, for example, an estimating department. The high-level goals and objectives of the organization might involve safety, profitability, employee satisfaction, and customer satisfaction. The estimating department touches each of these organization goals and should develop strategies and tactics that enable the achievement of these goals and objectives. Using safety as an example, one might put a strategy in place that says: (1) the estimating team will function in a way that provides physical and mental safety for our team members; and (2) the estimating team will estimate jobs in a way that allows for the construction team to execute the schedule in a safe manner. We would then need to operationalize these strategies with tactics and day-to-day activities. For example, we might schedule a quarterly mental health check in with the members of the estimating team as one important tactic.

> Consider the construction team of a new construction project. The team pursued that project with certain strategies in mind to achieve success such as early material buy-out, strong team communication, a close relationship with our client team, and a relentless focus on safety. Those strategies are the elements of the plan that help us reach the organization's goal: a safe, profitable project with a happy client. Those strategic elements then need to be operationalized into the daily tactics and activities that support each of these aims. For example, to develop a close relationship with our client, we might make sure to schedule a weekly high-level touch base to address any concerns, or we might make sure to walk the jobsite with them twice a week just to chat; or we might periodically survey the client team on their level of satisfaction. These are all examples of the many tactics that might support that strategy.

Your organization's ability to develop and execute the tactics necessary to enact the strategies that drive performance is essential to the success of you and your team. Tactical plans must be reviewed frequently and ensure that the behaviors of your team members are focused on the activities that drive the successful execution of top priority strategies.

> **Activity:**
>
> - Reflect upon the documented tactics of your team. Do your team members buy into these tactics? Is your team successful in implementing these tactics?
> - What additional tactics might be important for your team to create and document?
> - Is your team reliable in executing these tactics? Do people feel accountable for action? What might you do as a leader to achieve greater reliability?

Conclusion

Before he retired as Associate Director of Global Capital Management for Procter & Gamble, Mike Staun developed a high-level strategy of employing Lean construction practices on all construction projects to help their department and the organization achieve their goals and objectives. Several tactics were developed to enact that high-level strategy, including selecting six Lean practices to focus on; hiring management consulting firms (including ours) to help guide the construction project teams in the implementation of these Lean practices; and the creation of five-day collaborative project design meetings to execute Lean tactics. This process worked because the tactics were well-thought-out, connected to the high-level strategy, and aligned with the goals and measurable objectives of the teams and the organization. Adjustments were made along the way, but the plan was initially well-considered and reflected the diverse thinking of his team.

There is, of course, a tension between discipline and the agility that I believe is critical to sustained success. We need to have a plan, and we need to be able to turn on a dime when that plan is not working. More thoughts on this in Chapter 21!

References

Bossidy, L., Charan, R., and Burck, C. (2011). *Execution: The discipline of getting things done.* New York: Random House Business Books.

Collins, J. (2001). *Good to great: Why some companies make the leap...and others don't.* New York: Harper Business.

Gagel, G. and Beer, M. (2019). *Michael Beer discusses his book Fit to Compete* [Podcast]. 11 November. Available at: https://open.spotify.com/episode/1hW94f2hlImKZvvvaRqtti [Accessed: 28 April 2024].

Gagel, G. and Meares, A. (2020). *Anna Meares discusses leadership and team lessons from being an Olympic athlete* [Podcast]. 17 February. Available at: https://open.spotify.com/episode/3k1RqB8K3aiLVG5lRgkRol [Accessed: 28 April 2024].

Reding, M. (2023). 'What is a PESTLE analysis?' *CPD Online College Knowledge Base,* 18 April. Available at: https://cpdonline.co.uk/knowledge-base/business/pestle-analysis/ [Accessed: 14 June 2024].

Sridharan, M. A. (2021). 'Star model,' *Think Insights,* 13 July. Available at: https://thinkinsights.net/strategy/star-model/ [Accessed: 14 June 2024].

'Strategy.' (2024). *Merriam-Webster Online Dictionary.* Available at: https://www.merriam-webster.com/dictionary/strategy [Accessed: 14 June 2024].

'Tactic.' (2024). *Merriam-Webster Online Dictionary.* Available at: https://www.merriam-webster.com/dictionary/tactic [Accessed: 14 June 2024].

Teoli, D., Sanvictores, T., and An, J. (2023). 'SWOT analysis,' *StatPearls,* 4 September. Available at: https://www.ncbi.nlm.nih.gov/books/NBK537302/ [Accessed: 14 June 2024].

19

Utilizing Team Norms and Social Contracts

Introduction

During a recent construction project visit, the project team was presenting their outstanding results to date—ahead of schedule, under budget, great safety record, and a happy client. I asked the team why they felt they were coming together so well as a team, and their response was that it was probably mostly luck. I would argue it had much more to do with the make-up of the team and the implementation of many of the strategies and tactics discussed in this chapter. It may not have been deliberate, but the mesh of skills and the team's culture probably brought about these results. Leadership comes from all members of a team, not just the leader—and effective leaders are considered to be members of the team.

> *"Don't forget leaders are still part of the team. I think that's also really important."*
> (Anna Meares, speaking to Gagel, 2020 Greatness Podcast)

One question I am frequently asked is why organizations focus so much attention and energy on the creation and performance of teams and how teams differ from groups of people. Teams of people must rely upon one another to accomplish the task at hand, and they are one of the most effective strategies for accomplishing the critical work of the organization.

> Friend and Director of Knowledge Teams International Robert Marshall shares: *"We've worked with a total of 845 teams in the knowledge sector, about five and a half thousand team members. We often get questions about why invest in building teams because there are high transactional costs in getting teams to work. My summary is that teams are the best organizational structure for sharing and*

> *generating tacit knowledge and the reason I think that's so important is effective teams enable a low power distance culture or climate. Everyone is pretty much equal in terms of their capacity to contribute ideas, throw in either half-baked or dumb ideas, or to ask questions and to encourage that informal flow of tacit knowledge between team members. I think the other thing that really is critical for that to work is having a high level of trust between members."*
>
> (Robert Marshall, speaking to Gagel, 2024 Greatness Podcast)

When I first started with FMI in 1994, "Project Partnering" was a hot new methodology for helping construction teams complete projects without litigation, safely, on time, under budget, and at the highest quality (Gans, Gagel McComb, and Wambsganns, 1996). This work afforded me the opportunity to work with and observe hundreds of teams, including the good, the bad, and the ugly. My leadership positions over the years and time spent on teams in the construction industry have taught me important lessons about what works and what does not, and my PhD studies added an additional layer of research on high-performing teams. Through all of this, I have learned that being an effective team leader and teammate is crucial to the success of an organization.

As you read through this chapter, reflect upon the teams you have participated in and led. What has differentiated the high- from the low-performing teams? What ideas from this chapter might be useful to you in increasing your teams' performance? How might you share these ideas with others on your team?

Defining Teams

A team is defined as "**1:** a number of persons associated together in work or activity: such as **a:** a group on one side (as in football or a debate); **b: CREW, GANG 2a:** two or more draft animals harnessed to the same vehicle or implement" ('Team,' 2024). I find it interesting that the word "team" originates from Middle and Old English words for offspring, lineage, and a group of draft animals harnessed together ('Team,' 2024). Teams ARE the workhorses of organizations, accomplishing most of the work in the world—especially in the construction industry. In today's VUCA (volatile, uncertain, complex, and ambiguous) environment, organizations must have "rapidly reconfigurable resources," an ability for teams to form, reform, and reform again to meet the changing needs of the business environment. I often think of airline crews (probably because I fly so much!) Even though each crew is typically comprised of different people, everyone understands their role and the crews come together quickly as a team to achieve the goal of a safe (and mostly on time) flight.

> *"I realized there were more and more times when I saw people who were working with other people but not in stable intact teams. The boundaries of their teams were fluid and porous. The teams were in some sense not constantly but often shifting, often having new members coming in and others leaving, often needing to pull in a resource for a particular need."*
>
> (Amy Edmondson, speaking to Gagel, 2019 Greatness Podcast)

High-Performing Teams

When my daughter Regan switched from playing elite ice hockey in high school to being recruited to row crew for the University of Kansas, she must have received 20 copies of the book *The Boys in the Boat* by Daniel James Brown (2013). Interestingly, for years, I have thought of high-performing teams as an eight-person rowing team.

> *"All were merged into one smoothly working machine; they were, in fact, a poem of motion, a symphony of swinging blades."*
>
> (Brown, 2013).

Hopefully you have experienced this feeling of working with a team of people that achieved outstanding results.

The great news is that there is extensive research on what makes teams high-performing. The Aristotle Project, initiated in 2012 by Google employee Abeer Dubey and recent Yale graduate Julia Rozovsky, is an important example (Duhigg, 2016). The project researchers studied decades of research and about 180 of Google's teams to determine what differentiated good teams from great teams (Duhigg, 2016). The *New York Times* article highlighting these results was probably the first time the concept of psychological safety was broadly shared with the world.

> The Google research team identified five factors, in order of importance, that have the greatest impact on team performance (Duhigg, 2016):
>
> 1) Psychological Safety – Feeling safe to raise an issue or be vulnerable
> 2) Dependability – The ability to count on your teammates
> 3) Structure and Clarity – Clear roles, goals, and plans
> 4) Meaning – Finding purpose in the work
> 5) Impact – The work is making a difference

I believe we intuitively understand what makes great teams tick. When I work with leaders, I often ask them to describe the elements of high-performing teams, and many of the same elements appear on their lists—trust, clear goals, clear roles, and responsibilities. Leveraging research that underscores this intuitive understanding of what makes teams high-performing is an important step in driving team and organizational performance.

> *"The biggest single differentiator between teams that work well and teams that struggle is the **level of trust** in the team. That may sound sensible and intuitive to many people, but people in technical areas tend to think that technical expertise is what's critical. It is important, but it's not the most important. What we have is real data with real teams to demonstrate that. Your most important differentiator is the **level of goodwill and cooperation** in a team, the capacity of team members to do things to fill in any gaps, to say, 'I'll contribute to the best of my ability', rather than saying, 'That's not my job', or 'I'm not responsible for that'. It's that sense of mutual responsibility and commitment to the team that enables people to do whatever's in the moment while the team is in action. The third one is the **level of clarity of purpose**. The team really needs to know what it is established for, what its goals are and how each piece goes together so that each member of the team can contribute. Probably the fourth one is the **level of knowledge sharing** in the team, the capacity to say, 'Can you explain why you've done that?', or 'I've got a problem with this, can you help me?' or 'Have you had any experience in dealing with this aspect?'. This creates rapid informal flow of communication between members without needing to go through the boss."*
>
> (Robert Marshall, speaking to Gagel, 2024 Greatness Podcast)

If all of this research exists, why are there so many dysfunctional, low-performing teams out there? I believe it is because while many of us intuitively understand the elements of a high-performing team and have access to this research, we do not put the work into dealing with often unpredictable and difficult-to-understand human behavior. Patrick Lencioni captured five of these team dysfunctions: (1) absence of trust; (2) fear of conflict; (3) lack of commitment; (4) avoidance of accountability; and (5) inattention to results (Lencioni, 2002). Leading teams is about helping teams avoid and overcome these dysfunctions, helping them achieve their goals, and celebrating their accomplishments. How we as leaders enable teams to win is a gift to that team. I believe two key questions for us as leaders are, "Are we investing in people and the skills they need to be effective teammates?" and "Are there consistent consequences for inappropriate team behavior?"

One day, I walked into the office of a global airline that was spending roughly $800 million in construction annually to meet with the corporate real estate and construction leader and their six directors who were responsible for this program. Three of these directors were fairly new to the organization, and this leader had hired us to help this team redefine their roles to better meet the needs of their internal strategic business stakeholders. I began the meeting by drawing Tuckman's "forming, storming, norming, and performing" model up on a whiteboard and asking each member of the team to come up, one at a time, and place an "x" where they thought they were on this model (Tuckman, 1965). This model defines the stages of teams over time as: (1) "forming" – what I call the "honeymoon phase," where everyone is really nice to one another; (2) "storming" – this is the phase where misaligned behaviors and ineffective communication cause friction and frustration within the team; (3) "norming" – in this phase, the team defines acceptable and unacceptable behaviors and drives home accountability to team norms; and (4) "performing" – the phase of high performance for the team (Tuckman, 1965) (Figure 19.1).

Figure 19.1   Tuckman Stages of Teams.

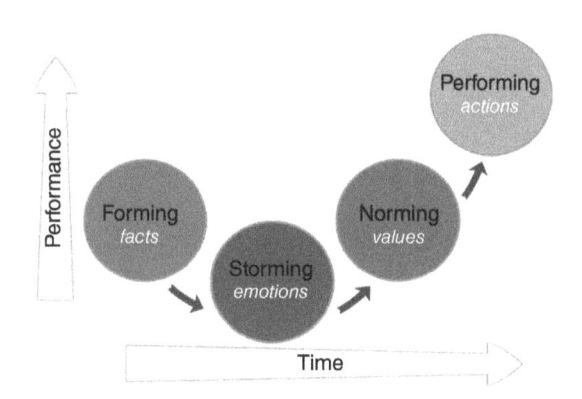

The first two directors who went up to the whiteboard were perhaps a bit cautious and drew their "x" up near the "norming" portion of the curve. Each subsequent person felt a bit braver and revealed the team's current level of dysfunction by placing their "x's" closer to the "storming" part of the curve. This five-minute exercise led to a candid and productive conversation about how this "storming" was manifesting itself with the team. We also discussed their aspirations as a high-performing team and what it would take to achieve that level of performance. I sensed relief that the team was able to talk about their frustrations in a constructive way that led them toward the high performance they all desired. These difficult but necessary conversations are all too infrequent, not because we do not care, but I believe because we are either too busy or do not have the tools to have the conversation.

Given my experience and research, I believe these are the key elements of high-performing teams:

1) **Deliberate Team Formation**

In the construction industry, we are constantly reforming teams for project execution and to accomplish support functions. How we go about assembling these teams is at times haphazard, which I find curious—especially given the amount of research available on what makes a great team. I will ask a project executive, "How was this team formed?" and they will respond, "Well, so-and-so was coming off of this project and so-and-so didn't want to travel…" I would argue that how we form teams and who we put on them should be more deliberate. Employing tools like a skills matrix is a great way to think about who to put on a team and starts with listing the skills needed for it to achieve success. You might think about people's DiSC profile, who is technically competent, who has process skills, and who has customer relation skills. It is also a great idea during team formation (and as team members enter and leave the team) to discuss the skills of each member so that people can leverage these skills.

> "*Many technical teams go straight to the task and spend little time on building relationships and getting to know one another, and this jumps up and bites them when they get to some sort of problem or crisis. One of the things that we encourage is teams investing some time right up front in getting to know one another and building respect for not only the expertise that each member brings but all of the other skills that can contribute to team success including research skills, communication skills, facilitation, recording, all of those skills that are really critical.*"
>
> (Robert Marshall, speaking to Gagel. 2024 Greatness Podcast)

Do you think the collective IQ of the team members dictates how smart the team is? If you said "no," you are right. In fact, Anita Woolley and Thomas W. Malone have conducted and replicated studies on team IQ and what matters most, and one of the most important factors is whether a team has women on it or not (Wooley and Malone, 2011).

> "*About half-way through the current database we started to collect information on gender composition, and what we found is that there is a positive correlation between the number of female members and overall team effectiveness. Five of the top seven teams in our database had more than 50% female members, and five of the lowest teams in our database had no female members. Four of the top seven teams had female leaders, and all of the seven lowest teams had male*

> *leaders. I think there are two probable explanations, and this is based more on observations and my interaction with teams rather than hard data. I think one of the effects of women on teams is that they reduce some of the macho decision making of men and encourage a more open and conciliatory, consultative climate. I think overall there's some data to suggest that in the population generally women tend to have slightly higher levels of emotional intelligence and concern for other members or empathy, and I think that probably plays out in this data."*
>
> (Robert Marshall, speaking to Gagel, 2024 Greatness Podcast)

Putting sufficient thought into the composition of each team is critical when striving for high-performing teams. Developing a process for doing so, providing thought leadership on great ways to consider who should be placed on a team and for what reasons, will set you apart as a leader.

Activity:

- Take time to reflect upon how you as a leader and your organization form teams.
- What could be done differently to be more deliberate in forming teams?
- How might you use a tool such as a skills matrix to help you understand the skills that are needed on the team and how each team member contributes?

---

2) Team Norms and Social Contracts

Thinking back to one of the multi-billion-dollar construction projects our team worked with to accelerate construction, we spent a great deal of time focused on our social contract; not just the important team values, but also the behaviors critical to successful execution of the project. One key behavior, collaboration, was difficult to emulate for some of our project leaders, and tough decisions were made to replace key leaders on certain sites. This sent a clear message that we were serious about our norms and social contract.

A norm is defined as "**2:** a principle of right action binding upon the members of a group and serving to guide, control, or regulate proper and acceptable behavior" ('Norm,' 2024). Norms are critical in that they define how we are going to behave with one another, and consistent behavior drives trust, reduces conflict, and enhances performance. A norm could be as simple as "we all agree to answer emails within 24 hours," and can head off communication challenges and frustrations. Norms are typically documents in some type of social contract that is discussed regularly within the team.

> Too often, we create or join teams without taking the time up front to identify our norms and document our social contract, even when we are trained to do so. During the leadership and ethics course I taught in the University of Denver's Online MBA program, I divided the students into working groups early in the course and asked them to develop a social contract of norms of behavior. About six weeks later, I remixed the students for one last project. After they delivered the project, I asked each group if they had taken the time up front to reestablish a social contract for this new project team. Some teams had, but most had not—and we just talked about its importance! Can you imagine a world where each team sat down, regularly, to discuss the acceptable and unacceptable behaviors of the team? Amazing.

Helping your teams document and adhere to a social contract of team norms creates a psychologically safe space to discuss behaviors and resolve conflicts. It is not enough to create the social contract of norms (Figure 19.2). The team must also frequently discuss the behaviors that support these norms, and not be afraid to have a difficult conversation when behaviors stray from the norms.

Team Social Contract Example

Figure 19.2 Example Team Social Contract.

- ✓ Disagree in the room, agree outside the room
- ✓ Listen to understand
- ✓ Respect the opinions of others
- ✓ Answer emails within 48 hours
- ✓ Advocate for each other when we aren't in the room
- ✓ Bring frustrations to that person, not others
- ✓ Respect people's time off

Activity:

- Do your teams have documented norms? If not, how might you go about helping them develop norms?
- How might you codify and share these norms in a social contract?
- What processes are in place to help teams celebrate behaviors that support the team norms? To deal with situations when people are not adhering to the team norms?

3) Team Performance Assessments and Continuous Improvement

It is also important that teams utilize feedback mechanisms to understand how the team is performing relative to the values, norms, goals, and objectives defined by the team. One great idea is to create a baseline for a team, either when the team is formed or when the team decides to work on certain aspects of the team's behaviors and performance. When we go to the doctor, she starts with assessment, including what the symptoms are and what tests we might need. We also typically start with the desired state of health (weight, blood pressure, etc.); the "high-performing" state of our body. This approach works with teams as well.

> *"It's not just the leader's responsibility. Yes, they do have positional power and they might have access to resources. However, high-performing teams don't just wait for the team leader to call out that behavior. Everyone is a leader in a high-performing team."*
>
> (Jan Terkelsen and Michelle Terkelsen, speaking to Gagel, 2020 Greatness Podcast)

You might try this exercise with your team:

- Ask the team to list the elements of a high-performing team (such as clear goals, high levels of trust, and respect) and explore the team's desire to perform at the highest possible level via a candid conversation on the topic.
- Baseline the team's performance on each element on the list via an anonymous survey.
- Meet with the team to review the results.
- Pick one low-scoring area to work on and identify three steps the team could take to improve.
- Create a plan and assign accountability to actions.
- Retake the survey in three-to-six months to measure improvement, and potentially pick another area to work on.

Figure 19.3 is an example of an engineering department client that performed this exercise and improved over time. Once a team defines the behaviors and parameters of success and is effective in measuring performance, efforts to improve as a team can be focused and effective.

> *"We encourage teams to take time periodically to review how the team is doing by carving out time at each meeting to reflect upon how we can improve the way that*

> *we work. The assessment process enables teams to evaluate and reflect on their results, and to think about what they can do to keep doing things well and improve things where they may be struggling. In fact, that becomes a lead indicator of team effectiveness and people can do something about the team relatively early in the project life cycle rather than wait till the end when it's too late."*
>
> (Robert Marshall, speaking to Gagel, 2024 Greatness Podcast)

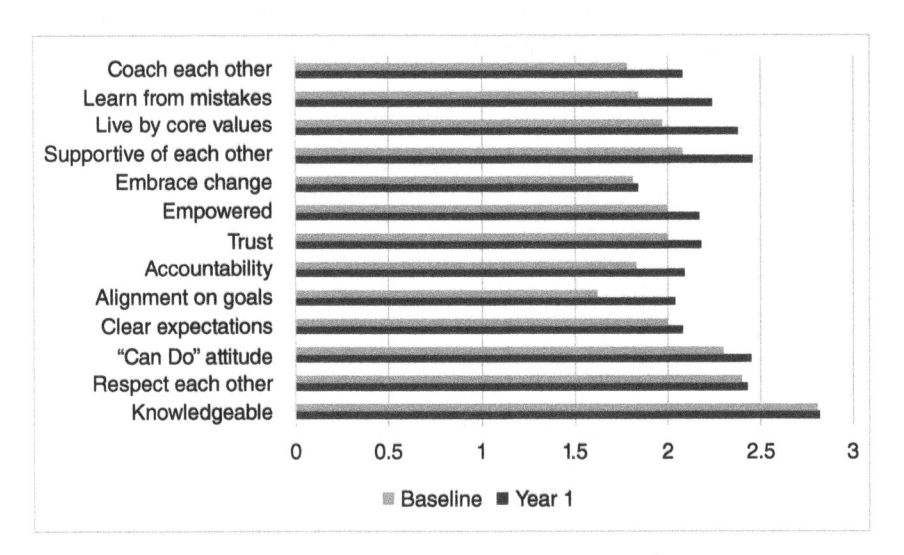

Figure 19.3 Engineering Team – Measuring High Performance, Baseline, and Year One.

Bringing someone outside of the team or organization in to facilitate the discussion can be helpful. Helping a team decide on the important metrics, baseline their performance on these metrics, focus on improving in certain areas, and then remeasure is an effective strategy for driving team performance.

Activity:

- How well does either the team you lead or the team you are a member of measure team performance in relation to the identified team norms?
- What might be done differently to improve the process?

4) The Impact of Organization Culture on Teams

I was hired by a global engineering firm to work with one of their teams that was experiencing a great deal of dysfunction that was causing client problems and employee turnover. I met with the five team leaders who agreed that yes, they were clearly in the "storming" phase of team development. I asked these team leaders to define the norms that would be important to achieving high performance, and then we picked one norm to work on for 30 days: that they would not talk negatively about each other behind their backs.

I flew back to meet with these five team leaders one month later and asked how well the team had done with this norm. The group sheepishly admitted that they were indeed still talking about one another negatively behind each other's back. I then realized that the culture of this client's office and the office leaders did not hold people accountable for poor behavior. Without any negative consequences (such as peer pressure or coaching by leaders), these poor behaviors were never going to change. I called my client and suggested they fire me until these deeper office cultural issues were resolved. The client was grateful for my suggestion, terminated our agreement for this team, and worked to address the issue.

> I once worked with a senior leader who had just joined an organization. They described the organization's culture and the corresponding impact upon them as being in an aquarium of murky water that causes team dysfunction. That person's comment, "*I need to leave before I forget how to swim in clear water.*"

It is imperative that leaders understand how the organization's culture is either promoting or discouraging high-performing team behaviors. No team operates in a bubble, and the context that the team operates within has a tremendous impact on a team's ability to achieve greatness.

> **Activity:**
>
> - How does your organization's culture support the elements of high-performing teams? How might that be reinforced and celebrated?
> - How might your organization's culture be undermining team performance? How might this be mitigated?

5) Effective Team Meetings

Our team consulted for a large client that was experiencing significant challenges in delivering construction projects. We created a detailed fishbone diagram of everything that "slowed them down" and tackled each root cause in turn. One significant issue the group identified was the number of meetings everyone was attending. In digging deeper, we realized that people were attending meetings they did not need to attend because the meeting's agenda was unclear. Because people had so many meetings on their calendar, they were frequently late to meetings, which held up important conversations. We baselined the amount of time people felt they were wasting in unnecessary meetings, and it was a large percentage of people's time. I am certain you can relate, as I often hear people articulate their frustration with ineffective meetings.

We then pulled together a working group to define the meeting protocols for the team:

> - Every meeting has a pre-published agenda and established purpose.
> - Every meeting starts and ends on time.
> - Every meeting has a meeting "owner," the person responsible and account-able for keeping the meeting on track.
> - Every meeting has a "parking lot" to record important thoughts and issues that are not relevant to this meeting but should be discussed at a future meeting.
> - Every meeting concludes by reviewing agreements and actions, along with any relevant future meetings that need to occur.

These meeting rules were posted in every meeting room in the department, but I knew that was not enough to significantly change behavior. I sat down with the team's leader and asked him to walk into one meeting unannounced every day for a month and ask them one question, such as *"Was the agenda sent out ahead of time?"*, *"Did this meeting start on time?"*, *"Who is the meeting owner?"*, or *"Where is the meeting parking lot?"* After about three weeks of this, everyone had received the message that we were serious about improving the effectiveness of meetings, and the team began to implement the meeting protocols. It worked. Three months after the baseline, we remeasured the amount of time wasted in meetings, and it had gone down by the equivalent of eight full-time employees.

> Teams pay attention to what leaders ask about. If you are serious about making a change, ask if it is happening, often.

My team worked with a big-box retail company to improve their store construction speed, and the organization's culture seemed to reward people and rate their importance by the number of meetings they attended a week. I suggested that the leader of construction ban meetings for a month! I believe wasted time in bad meetings is not only impacting productivity, but it is also a culture killer because these meetings are demoralizing. People and teams do not have time to accomplish their work due to the hours spent in ineffective meetings.

There are many experts out there with more advice on how to solve this meeting problem, one of them being my friend Donna McGeorge, author of *The 25 Minute Meeting* (McGeorge, 2018). I encourage you to examine the effectiveness of your team(s) meetings and take the steps necessary to improve effectiveness.

Activity:

- How effective are your meetings? What norms might you put in place to improve the effectiveness of your meetings?
- During your team's meetings, think about the way people's behaviors reflect the culture of your organization.
- What is tacitly being conveyed as "okay behavior" by people's actions? Is it okay to tune out while someone is speaking, or are we respecting each other's opinions? Are we talking over one another? Is there shouting? What does this say about our value of respect, for example?

Conclusion

Early in my management consulting career, I was asked to work with a construction project team that was, as they described, "blowing up." The team was behind schedule, over budget, and it was still early in the project. The leadership of the team turned the situation around by: (1) recognizing the poor results early; (2) asking for help early; and (3) committing to replacing team members who did not respond to requests to change their behaviors. These swift and appropriate actions helped the team come together as a high-performing team that achieved outstanding results.

Author and mindfulness expert Michael Chaskalson shares, *"First spoken of by my colleague here in the UK, Professor Michael West, team mindfulness centers upon a team becoming aware of itself. This isn't about individual members of the team practicing mindfulness. It's about the team itself becoming self-aware of itself as a team. The team turns its attention collectively to things like its tasks, its goals, how it's doing. Importantly, people turn their attention to the dynamics within the team. How are we doing together? How are we with each other? What's happening here? How are we all doing? We create a culture where people can bring their whole self to work and that for me is one of the big takeaways of team mindfulness, that it creates an atmosphere of psychological safety."*

(Michael Chaskalson, speaking to Gagel, 2020 Greatness Podcast)

Whether leading or a member of a team, you are responsible for your behavior and how you contribute to the team's performance. Your ability to understand the elements of high-performing teams, politely and appropriately call out bad behavior, and help your teammates thrive with the skills they need to be successful on the team will be respected and appreciated.

> *"One of the indicators we have found with highly effective teams is that people say that it is fun to be a part of the team and they actually look forward to work. There are still the challenges, the issues, the obstacles. However, there is this sense of collaboration and teamwork and having fun in amongst that."*
>
> (Jan Terkelsen and Michelle Terkelsen, speaking to Gagel, 2020 Greatness Podcast)

References

Brown, D. J. (2013). *The boys in the boat: Nine Americans and their epic quest for gold at the 1936 Berlin Olympics*. London: Penguin Books.

Duhigg, C. (2016). 'What Google learned from its quest to build the perfect team,' *The New York Times*, 25 February. Available at: https://www.nytimes.com/2016/02/28/magazine/what-google-learned-from-its-quest-to-build-the-perfect-team.html [Accessed: 11 June 2024].

Gagel, G. and Chaskalson, M. (2020). *Michael Chaskalson discusses team mindfulness* [Podcast]. 23 November. Available at: https://open.spotify.com/episode/7eFjqPm7HrcMvR77x8K1rp [Accessed: 28 April 2024].

Gagel, G. and Edmondson, A. (2019). *Amy Edmondson discusses her book The Fearless Organization on psychological safety* [Podcast]. 30 October. Available at: https://open.spotify.com/episode/6joytIVhV3XlGFZx6rcodC [Accessed: 28 April 2024].

Gagel, G. and Marshall, R. (2024). *Robert Marshall discusses his research of over 800 teams* [Podcast]. 3 May. Available at: https://open.spotify.com/episode/5SsVvBeUjzCZ3GxWSiABO9 [Accessed: 14 June 2024].

Gagel, G. and Meares, A. (2020). *Anna Meares discusses leadership and team lessons from being an Olympic athlete* [Podcast]. 17 February. Available at: https://open.spotify.com/episode/3k1RqB8K3aiLVG5lRgkRol [Accessed: 28 April 2024].

Gagel, G., Terkelsen, J., and Terkelsen, M. (2020). *Jan and Michelle Terkelsen discuss creating high performance teams* [Podcast]. 3 February. Available at: https://open.spotify.com/episode/4U22Jcl7IYP2Ana37l4zrI [Accessed: 28 April 2024].

Gans, J., Gagel McComb, G., and Wambsganns, E. (1996). 'One partnering success secret: Set high goals,' in Godfrey, K. A. (ed.) *Partnering in design and construction.* New York: McGraw-Hill, pp. 75–87.

Lencioni, P. (2002). *The five dysfunctions of a team: A leadership fable.* San Francisco: Jossey-Bass.

McGeorge, D. (2018). *The 25-minute meeting: Half the time, double the impact.* Hoboken, NJ: John Wiley & Sons.

'Norm,' (2024). *Merriam-Webster Online Dictionary.* Available at: https://www.merriam-webster.com/dictionary/norm [Accessed: 14 June 2024].

'Team,' (2024). *Merriam-Webster Online Dictionary.* Available at: https://www.merriam-webster.com/dictionary/team [Accessed: 14 June 2024].

Tuckman, B. (1965). 'Developmental sequence in small groups,' *Psychological Bulletin*, 63(6), 384–399. Available at: https://web.mit.edu/curhan/www/docs/Articles/15341_Readings/Group_Dynamics/Tuckman_1965_Developmental_sequence_in_small_groups.pdf.

Wooley, A. and Malone, T. W. (2011). 'Defend your research: What makes a team smarter? More women,' *Harvard Business Review,* June. Available at: https://hbr.org/2011/06/defend-your-research-what-makes-a-team-smarter-more-women [Accessed: 14 June 2024].

20

Attracting, Hiring, and Retaining Diverse Talent

Introduction

Talent shortages and population demographics are causing organizations to think more strategically about how to attract, hire, and retain top talent. As the leader of a team or organization, you are only as successful as the talent you deploy to successfully execute your strategies and tactics. This is not just an organizational challenge; it is an industry challenge. Characteristics of the construction industry (such as pressure and stress) limit our available talent pool. I hear people say, "They just needed a less stressful job," and I ask, "Why don't we make their job less stressful?! Why are we okay being the industry with the highest suicide rate?" As an industry, we can do a better job attracting, hiring, and retaining diverse talent, and it begins with each team and organization in the industry. This is critical to our industry's sustainability.

> *"I went to the caucus around 2016. At that time, we were a $200 million company and by 2020 we were a $500 million company; and we had started a self-perform business from zero to $45 million during that period as well. We had profits that were two-and-a-half times better after adjusting for inflation than we've had in our 76-year history. Is it all because of diversity? Of course not. But a significant piece of it is, and my grounding for that would be that in creating the inclusive culture, it's become more engaging. We did a Crain's Cool Places to Work survey and the results blew me away. We were at 99 where others had said achieving over 45 was difficult. There were three questions that everybody in the company said, 'yes' to and they were, 'I'm proud to work for this organization', 'I'd recommend this organization's products or services to a friend', and 'I would recommend that a friend come to work here'. When you have a high level of engagement, that generally leads to people taking care of our client which leads to return business. It's increased the talent level. It's increased the feeling of belonging and safety for*

Building Women Leaders: A Blueprint for Women Thriving in Construction,
First Edition. Gretchen Gagel.
© 2025 John Wiley & Sons, Inc. Published 2025 by John Wiley & Sons, Inc.

> *everybody in the company and that creates positive results, and so there's a lot of talented people here doing a lot of good things."*
>
> (Crain's Chicago Business, 2023; Sam Clark, speaking to Gagel, 2023 Greatness Podcast)

As you read this chapter, consider the strengths (and perhaps the challenges) of your organization related to attracting, hiring, and retaining diverse talent. Is your organization committed to inclusion and creating the best possible working environment for your people? How well are you measuring your success in these areas? Is your organization leveraging creative thinking in overcoming our industry challenges to develop a highly engaged workforce?

Attracting Diverse Talent

Years ago, I heard that a construction association in the southern US was recruiting high school girls and their often low-income moms to join the trades together, so that these two women would experience a built-in buddy system. This is a remarkable example of someone thinking creatively about how to attract diverse talent to our industry. Many men and women I interviewed for this book stated that they kind of "fell into construction," or that they thought it was going to be a "short-term job until they found something better." The construction industry was not their first choice. We have, in general, an image problem where we are not viewed as an industry of choice, and this is a challenge all industry leaders need to work toward overcoming. This pains me because we are such a remarkable industry, and I want the world to know how gratifying a construction industry career can be.

Here are my thoughts on how we, as industry leaders, can help attract diverse talent:

- **Speak about the many benefits of our industry.** I encourage everyone to share the pride we have in all that we do to build and maintain the assets of society. Tell the stories of the projects we build, the infrastructure we put in place, and how we benefit society by ensuring that buildings withstand earthquakes, people have safe transportation, and we have homes to live in and offices to work in.

  When I first returned to the construction industry in 2013, most of my Denver friends were unaware of my past career in the industry. "Construction? You work in construction? Why construction?" I enjoyed responding with the many significant projects that have improved their lives—the new lightrail system or tunnel under the Sydney Harbour, the data centers that hold all that "stuff in the clouds" they are viewing on TikTok, the playgrounds their children play on.

Telling these stories with pride helps improve our industry image, and I still think we need a hit television series!

- **Expand our recruiting to include diverse sources.** Years ago, I heard the male CEO of an energy company speak on a panel at a women in energy conference. This CEO told his recruiting people that 10% female engineer applicants was not good enough, and that he did not care how many universities they had to visit, he wanted 50%. We as leaders need to be that demanding in ensuring a diverse talent pool to draw upon. *"Treat them the same as you would treat a man. Hire the right person, cast a wide net, and interview candidates based on their skills. Don't hire someone just because she is a woman."* (Petrillo, 2023).

> *"We set goals for college recruits as our pipeline for future leaders comes primarily from our recruits. We are more intentional and continue to advance their careers. We are transparent on everything we do and think, do we have a diverse project team? We need 3 million people to make the industry sustainable; it's our responsibility to industry and society. We should be motivated to do the right thing for people and it makes our business successful."*
>
> (Reilly, 2023)

- **Deliberately recruit from outside of our industry.** In 2023, I had the chance to meet two extraordinary women: Christa Andresky, Executive Vice President and Chief Financial Officer of Turner Construction Company, and Beth Soukup, Chief Financial Officer of JE Dunn Construction. Interestingly, both were aggressively pursued for their roles from outside of the industry, and I believe these women exemplify out-of-the-box thinking on how we draw people (especially women) into our industry.

> *"I had no intention of leaving public accounting for a career in the construction industry, and in fact declined this CEO's request for an interview until I was finally convinced to meet with him. I had no idea how fulfilling my work as CFO of JE Dunn would be, and I feel I've brought a fresh perspective to my business partners here and the industry."*
>
> (Soukup, 2023)

Andre Noonan, COO of ACCIONA Australia, spoke at a NAWIC Australia conference about how he challenged his team to achieve 50% women on a construction project at a girls school in Melbourne. When they could not find women with certain skills, they trained them: *"One of the key aspects of the project was very technical drilling, deep holes through complex geology and a high-water table in Melbourne with water egress challenges. We didn't have any female drilling rig operators; and I said, 'Well, we've got four weeks. I learnt 30 years ago to operate a drilling rig in a week and was sent out to site and then learnt on the tool for probably two or three years after that. You've got to begin somewhere so there's your challenge, you've got four weeks.' The trainer was an in-house experienced male operator who I've been working alongside for decades. We worked through it, and we trained women. The microcosm of the drilling industry within the building industry, you can't stress how blokey and strong that is, so to bring in a female piling operator, I'd almost guarantee it's a first in Australia. It's a bit like you read about Rio Tinto and BHP with the big heavy haul, 300-ton dump trucks up in the Pilbara. They are finding their maintenance and safety with the female drivers is 30 or 40% less than with the males"* (Andre Noonan, speaking to Gagel, 2022 Greatness Podcast).

- **Share success stories of attracting diverse people.** We need to tell the stories of diverse people being successful in our industry so that more people envision themselves as belonging. We need to tell the stories of leaders who demand (and achieve) diversity and inclusion on their projects.

    "It just so happened that both of my daughters went to this school, but it was a little bit coincidental. I just felt, it being girls' school under construction and we're talking about gender diversity, we're able to make these decisions, and we said, 'Why not shoot for 50% women?' I made the call and then we had a couple of internal discussions, and initially there was a lot of resistance against it. I just got up from the meeting and said, 'That's it. We're going for 50/50, the only floor is fifty percent, and I don't care how high you go in terms of female participation. And by the way you've still got to fulfill all your other obligations as general managers or construction managers and do it on time and on budget.' It needed someone just to make the call.

    I presume I left the room and there was a lot of white board sessions and 'how are we going to do it.' To everyone's credit, we did it, and it was incredible, and we were able to achieve the budget and the program and were able to bring new participants to the industry. It was very important to make sure we brought the entire organization along on the journey; and I must say I was quite nervous giving that sort of unilateral direction, but I did. You know that adage, you've got to break some eggs to make an omelette, and I think to make this omelette it needed just that, 'No, we're doing it,' and away we went, and we did it. It's no credit to me. It's a credit to the team and the client, everyone. I think we ended up with 53% overall."

    (Andre Noonan, speaking to Gagel, 2022 Greatness Podcast).

- **Be more intentional about taking chances on people**
    I hear people speak as an industry about the lack of talent out there, but are we really considering all the types of people who might fill a role? Are we taking

chances on someone that maybe does not have the exact work experience we require, but instead has many skills that are transferable to a new job situation?

> *"As an industry we need to take a chance more often, understand that if you have a female leader, a project executive, the female will deal with problems a little differently. 'I'm not sure she can handle the stress', 'she doesn't smile as much', these are not substantive comments. We as an industry need to be more intentional about hiring and promoting to force the issue."*
>
> (Andresky, 2023).

These strategies and tactics are critical to our industry's sustainability. You may not oversee recruiting strategies, but I believe everyone within an organization can influence recruiting strategies to enhance the diversity of our industry.

Activity:

- How do you personally tell the story of how great our industry is to those outside of it?
- What is our organization currently doing to attract diverse talent?
- What more can be done? How can you as a leader influence these strategies and tactics?
- What is one out-of-the-box idea you could suggest for attracting more people into our industry?

Hiring Diverse Talent into the Right Roles

When I was CEO/President of Continuum Advisory Group, we could have reached out to find people with management consulting and construction industry experience. Instead, we looked for people with intellectual curiosity who were strong collaborators with the right ethical mindset. We could teach them management consulting, and we could teach them the construction industry, but we did not feel we could teach them to be ethical or curious.

My philosophy in life has been to surround myself with people who are smarter and more talented than me, and then take full advantage of their brilliance. Hiring the most talented people means finding people with the specific skills you need who are a great cultural fit because their values align with those of the organization. I encourage you to think deeply about the question of what skills, behaviors, and values are necessary for a position. How do you develop this list? Frankly, some of this is through trial and error. At one company, we hired a person we felt ticked all the boxes for skills and values. However, we soon realized this person was not adept at listening to a group discussion, mentally processing the nuances of what was being said, and quickly capturing these ideas on a flipchart. This is management consulting 101, and we had completely missed it.

This example points out the challenge of how we measure skills, behaviors, and values. How do you measure ethics? How do you measure the ability to quickly and concisely capture ideas on a flipchart? I have found role plays to be one important tool. For example, in the MBA ethics and leadership course I taught, I asked students to anonymously respond to the following situation: "You are at a restaurant having a nice meal and bottle of wine with your partner, and when the bill arrives, the waiter has accidentally left off the bottle of wine. What do you do?" Their answers would provide insight into their moral compass. Some students said it would depend on whether it was a mom-and-pop restaurant or a large chain, or on how good the service was, or on how much money they made. My answer: you call the waiter over and tell them they left the bottle of wine off the bill! This question captures a real situation that happened to my partner and me, and I was shocked by how surprised and grateful the waiter was when I pointed out his mistake, stating that the cost of the wine would have come out of his pocket. Profiles such as DiSC also provide insight into people's skills and behaviors.

> When I was President of a company, our team was hiring a new CFO. We narrowed the candidates for the role to two highly qualified individuals, who we then asked to take the DiSC profile assessment. Individual A was high in "I," very focused on people and collaboration, while Individual B was high in "C," attention to detail and correctness. While everyone on the team really liked Individual A (of course, they were the people person!), we hired Individual B because the role required accuracy, and this person was highly successful.

I believe we also need to think creatively about the roles we are hiring people into. I find organizations to be rigid in their descriptions of certain roles—project engineers do this, project managers do this, and these are the hours that are required. I approached one of my managers about cutting back on my travel schedule. I had thought that once the kids were out of diapers and sleeping through the night, the travel would become easier. Instead, I found it more difficult because as they started to play sports and have school plays, they wanted mom there even more. What I discovered was a lack of creativity in thinking about our roles within the organization and how one was rewarded for performance.

> When I returned to the construction industry in 2013, I was determined that our team would have more flexibility in how we thought of people and their

roles. Our leadership team defined each person's role differently according to how much time they were to spend on marketing, business development, sales, consulting, project management, and client management. We focused on leveraging the unique skills of each person while ensuring that all of these roles were adequately covered within the organization. We implemented a time tracking system to hold people accountable to these roles and make necessary adjustments in people's time allocations. This was a unique way to approach our work, with everyone having slightly different roles.

I believe that great leaders creatively solve the challenges associated with hiring the right people into the right roles by effectively defining the roles and having fluidity in how roles meld together to achieve the activities of a team. Great leaders are adept at determining the skills and behaviors necessary for success in a role, and well-versed in finding ways to measure these skills, behaviors, and values in job candidates to ensure that the right people are brought onto the team.

Activity:

- What values are you looking for in people for your team?
- Do you have an inventory of the skills and behaviors necessary for people on your team? If not, what one step might you take to improve this?
- Does your team understand the roles that must be accomplished for the team to be successful?
- How flexible are you in how these roles and responsibilities are assigned to leverage the skills and behaviors of team members?

Keeping the Right People

Yes, the world has become more mobile, and the days of most of your workforce retiring with a gold watch after 50 years are behind us. But there are organizations out there with lower turnover rates and strategies and tactics that keep people. The care and nurturing of your people to retain your top talent are critical.

"We would have about 10% of the company turnover each year and that's high historically for us, and after implementing our inclusive culture, turnover went down to losing 2-to-3 people a year from our staff of 200."

(Sam Clark, speaking to Gagel, 2023 Greatness Podcast)

Here are my suggestions on how to retain your diverse talent:

- **Keep your finger on the pulse of how your people feel they are treated.** Do they feel supported? Do they have the resources they need to be successful? Are they treated with respect? Do we care about their careers and aspirations? Are we looking out for them when they face personal issues? Often, I hear leaders say they did not realize there was a problem with one of their team leaders, or that they did not realize how unhappy someone was in their role until they left. While I love the construction industry, it can be a stressful, challenging environment for our people. Ensuring that your people are engaged and feel safe in raising concerns is of the utmost importance. How do you do this? That "boots on the ground" leadership I spoke of earlier, being out with the team, anonymous surveys, strong whistleblower policies, sharing examples of people raising an issue and thanking them, and effectively dealing with an issue when they raise it.
- **Let go of people (and especially leaders) who do not fit your culture.** Years ago, I was catching up with a CEO and long-time friend whom I had not seen in years. This person was about three years into his role as CEO and working hard to shift the culture of an organization that was not as agile as it needed to be. He shared his frustration with the pace of change, and I asked him, "Who have you fired?" He was surprised by my question. I explained to him that if the cultural change is not occurring, then there are people who are not getting on board. If those people remain, not only do they send a signal that we are not serious about changing your culture, but they are probably also exhibiting behaviors that do not align with your values, and that may be causing people to leave. It harkens back to my story of Bill Boyar, Founding Shareholder of BoyarMiller, and the hard decision his firm made to fire one of their top producing people.
- **Help your people deal with the complexity of today's business environment.** In his book *The Patient Will See You Now,* Eric Topol argues that we have not seen a revolution in the availability of information caused by the internet since the invention of the printing press, and that the speed of technology evolution is outstripping people's capacity to adapt (Topol, 2016). I agree. When I was earning my MBA, there were no personal computers, no cell phones, no emails, and no online textbooks. I recently attended an International Women's Forum (IWF) conference in Helsinki with an outstanding panel on quantum computing, and the next phase of technology adoption (including artificial intelligence) is mind-blowing. Helping our talented people cope with the complexity of the world is a critical step to retaining them.

We keep our top talent by ensuring that they feel well treated—a "whole human" approach of caring about their mental and physical wellbeing and their fulfillment at work. We keep our top talent by letting other people go—people who are not a great fit with the values of our organization. We keep our top talent by helping our people cope with the complexity of the world via strong internal processes, effective technology adoption, and a strong support system. Your organization cannot afford to lose great talent, and your ability as a leader to enact these important concepts is vital.

Activity:

- Is your team frequently losing great people? Why? What is the root cause of their departures, and how can you address this?
- How effective are you as a leader in keeping your finger on the pulse of how people on your team feel about you, the team, and your organization? What might you do differently?
- How are your people coping with the complexity of the world? Is this a topic of discussion?

Selecting the Right Managers and Leaders

In the engineering and construction fields, why do we repeatedly promote really good engineers to management and leadership positions, often without related training? I am a big proponent of having technical career paths and management/leadership career paths within organizations, as well as for organizations to reward people for being both great technical subject matter experts and great managers and leaders. Selecting the right managers and leaders starts with thinking about the skills that person needs to be successful in their role (not just the technical skills, but the people and relationship-building skills) and then selecting someone who has already demonstrated these skills or you believe has the capacity to learn these skills. Then you invest in these leaders, you coach them, you nurture them as managers and leaders, you set clear direction, you hold them accountable, you celebrate their success, and you coach them to higher levels of performance.

During my time teaching at the Australian National University, I was exposed to the outstanding research of Dr. Michael Platow on social identity theory and leadership (Haslam, Reicher, and Platow, 2020). This research helps us understand unconscious bias and how groups of people relate to one another and their leaders, including the fact that teams accept leaders who match their perception of the characteristics of a leader that fits within their group. I vividly remember one research study Michael shared. If people wearing a sweatshirt from your university say the popcorn is good as you walk into a movie, you will eat more popcorn. If a person wearing a sweatshirt from a different university says the popcorn is good, you won't. Crazy! Michael has replicated this research over and over, uncovering the unconscious bias and in-group/out-group thinking that happens subconsciously. I do not think that means we cannot hire and promote leaders who may not match the perceptions of leadership of their teams. What we DO need to do is be present to the unconscious bias we all have when selecting leaders, and we need to help diverse people thrive in settings where they may not look like those they are leading.

> *"We need to pay attention to what we call psychological groups. Of course, if everybody's a member of the same organization that organization is what we might call the sociological group. It has an organizational structure. It has rules, norms, hierarchy. But leaders must work at constructing this shared sense of a collective and there are three things that anybody in a leadership role could do. The first one is to show the people you want to follow you that you respect them. Listen to their views, ask them their opinions on things. The second thing you could do is be fair. You need to be seen as being fair and you've got to listen to their views. We call it giving voice. It's got to be more than what we call non-instrumental voice. You've got to give people voice and you've got to say, 'Look, this is something that we may be able to implement,' you give it credit because otherwise people just think you're giving them empty voice. The last one is to trust people. If you say to your employees, 'Take this project and run with it and I'm going to trust you to get it right,' they may get it wrong. But you as the manager may get it wrong too. If you trust people, you're really going to signal to them that you and they are all on the same team. You will be able to then influence people and when I say influence, I mean actually say, 'I think this is the way we should go,' and people will say, 'Yes, I think that's the way we should go' without you having to pay them more, without having to give a threat, without having to expend extra resources on monitoring and so on. They'll want to do it because they see themselves as part of the bigger collective."*
>
> (Michael Platow, speaking to Gagel, 2019 Greatness Podcast)

We also need to be intentionally promoting diverse leaders. You cannot be what you cannot see. As Jane Beaudry, HSE Director Life Sciences North America for Jacobs, shares, *"On the trade side I don't see similar intention. I have never seen a female foreman, rarely a female supervisor. Women entering the trades don't see role models"* (Beaudry, 2023).

How you go about selecting and developing managers and leaders in your organization is a critical step in retaining these people through advancement. Providing them with the strong relationships building skills results in the retention of those they lead. We cannot over-invest in the management and leadership skills of every person in our organizations and especially those who lead others.

Activity:

- What process does your organization and/or team use to select managers and leaders?
- Is this process effective? How might it be improved?
- Are you adequately equipping new managers and leaders to manage and lead? What might be done to enhance their leadership skills and set them up for success as leaders?

Conclusion

During an association board meeting on the topic of retaining talent, one board member shared a story of how his organization was becoming more creative in meeting project staffing needs and the needs of their employees. For example, a new construction project started that required someone on the team to relocate for a few months. The most likely person for the project manager role was currently coaching their child in baseball. Rather than relocate this person and risk losing them, the company divided the project in half, sending a different person to manage the construction project for the first half, which allowed the person coaching their child to relocate later and not miss the season.

This is a great example of the flexibility we need to exhibit not only to keep great talent on our team, but also to attract the talent our industry so desperately needs. I believe the organizations that win the talent war will win the competition and profitability wars as well. Every person in every role influences how we attract, hire, and retain people. Changes will not happen overnight, but incremental improvement in these efforts can pay big dividends.

References

Andresky, C. (2023). Interview by Gretchen Gagel [Zoom]. 21 December.

Beaudry, J. (2023). Interview by Gretchen Gagel [Zoom]. 26 October.

Crain's Chicago Business. (2023). *How we compiled Crain's best places to work 2023*. 13 September. Available at: https://www.chicagobusiness.com/best-places-work/how-crains-best-places-work-2023-was-compiled [Accessed: 14 June 2024].

Gagel, G. and Clark, S. (2023). *Sam Clark discusses diversity equity and inclusion in the construction industry* [Podcast]. 21 April. Available at: https://open.spotify.com/episode/6pSdP91DlpLsBVX6OmtokB [Accessed: 2 May 2024].

Gagel, G. and Noonan, A. (2022). *Andre Noonan discusses gender diversity in construction* [Podcast]. 7 January. Available at: https://open.spotify.com/episode/5aM6gnz96oIbulicBcEI5Z [Accessed: 1 May 2024].

Gagel, G. and Platow, M. (2019). *Michael Platow discusses how leaders can create a psychological connection to their team* [Podcast]. 25 November. Available at: https://open.spotify.com/episode/1RRVFDM7CE0EzlKqiDHQ2D [Accessed: 28 April 2024].

Haslam, S. A., Reicher, S. D., and Platow, M. J. (2020). *The new psychology of leadership: Identity, influence and power*. 2nd edn. New York: Routledge.

Petrillo, K. S. (2023). Interview by Gretchen Gagel [Zoom]. 20 November.

Reilly, T. (2023). Interview by Gretchen Gagel [Zoom]. 25 October.

Soukup, B. (2023). Interview by Gretchen Gagel [Zoom]. 12 July.

Topol, E. (2016). *The patient will see you now: The future of medicine is in your hands*. New York: Basic Books.

21

Creating Agile Teams and Organizations

Introduction

When I was transferred to Beech-Nut Baby Foods, we had a "burning platform"—a metaphor derived by Daryl Conner to describe an issue that forces an organization to change (Conner, 2012). Our Beech-Nut leadership team was given 12 months to break even, or President Jim Nichols indicated that Ralson Purina would shut the company down. Even with that burning platform, change was difficult, and I remember Jim asking us all to read *Teaching the Elephant to Dance*, my first foray into studying theories of change management (Belasco, 1991).

But what if you do not have a "burning platform" to drive change? Or worse yet, what if you are Kodak or Blockbuster and have a "burning platform" but do not recognize it? I returned to study for my PhD in 2014 because I wanted to understand how leaders create nimble organizations that are highly capable of changing in the face of shifting business environments. The ability to create an agile culture whereby the entire organization understands that constant evolution is critical to maintain success is required of today's leaders, teams, and organizations. During my PhD dissertation research, I studied 126 business units housed within 47 corporations of more than 1,000 employees. I wanted to understand if certain behaviors of the leaders of these organizations facilitated greater agility for their teams. The good news is that after collecting over 10,000 pieces of data, I did in fact discover that there are certain leader behaviors that support agility, as well as some that do not.

> *"I argue that the single greatest habit we can build as leaders right now is becoming comfortable with being uncomfortable. No one can come on this podcast and tell you, Gretchen, that they definitely know what the future will look like. But one thing we can be pretty sure about is that the pace of change isn't slowing*

> *down. The uncertainty and complexity of the world we are leading in will only continue. We need to be thinking about how we get match fit for that; and that means getting comfortable with stepping outside our known world."*
>
> (Holly Ransom, speaking to Gagel, 2022 Greatness Podcast)

During my PhD studies, I revisited decades of traditional change management theory, including Lewin's seminal "unfreeze, change, freeze" change model, Prosci Founder Jeff Hiatt's ADKAR model, and John Kotter's eight-step model of change (Lewin, 1947; Kotter, 2012). These traditional change models have important features that should be incorporated into your leadership thinking while you "build your change muscle" as an organization. However, many recognize that these models are based on an idea that a team or organization experiences significant stretches of stability between times of turbulent change. The theory of organization agility recognizes that these times of stability are becoming shorter, and that change is more constant in the business environment (Worley, Williams, and Lawler, 2014; Meyer, 2015; Holbeche, 2023). For this reason, I have chosen to focus on organization agility in this book.

> *"Efficiency remains important, but the ability to adapt to complexity and continual change has become an imperative."*
>
> (McChrystal *et al.*, 2015)

As you read this chapter, consider your own personal level of comfort with change and the level of comfort with change of those you lead. Reflect upon how the information in this chapter might benefit you, your team, and your organization during your quest to remain relevant in an ever-evolving business context. You might also think about how to share this information with your team.

People and Change

Before we dive into the concept of organization agility, it is worth noting that organizations do not change; the people within them change. As leaders, our ability to influence the behavior of others—to convince others to change—is at times challenging. The DiSC profile is one tool that helps us understand why some people are naturally less comfortable with change. Having said that, for change and agility to occur within an organization, all people must change.

> **Mountain Change Model**
>
> Back in my early leadership days during the transformation of Beech-Nut Baby Food, I learned an invaluable lesson about people and change. I think of setting a strategy as setting the course for a boat. The leader says, "The boat is going here," as in, "We are implementing self-directed work teams," and there are people who immediately see the value of that change and jump into the boat. These are the people you work with for quick wins in implementing the change. Then there are people with one foot in the boat and one foot on the dock, and you can pull them into the boat as they see progress being made. There are people on the beach, and people in the grass near the beach, and you are working hard as a leader to move these people closer and into the boat.
>
> Then there are people standing on a mountain 50 miles away saying, "I am never getting into that boat (Figure 21.1)." As a young, somewhat egotistical manager, I thought, "I can get those people off the mountain." Guess what? In most instances, no, they are not coming off the mountain. More importantly, when you focus attention on them, people start getting back out of the boat because they are not certain you are serious about the change, and they think the mountain might be a safer place to be. If you keep the people on the mountain, it is incredibly detrimental to the culture and might derail change. Instead, as a leader, you need to coach these people very specifically for three-to-six months on the specific behaviors that need to change and ask them to "get on board"; and then graciously ask them to find a "new boat" in a new organization rather than passing them off to another team, as often happens. This takes courage, but it is well worth the effort. *"Sometimes there are people that are never going to come on board, they're never going to put their oar in the water at the same time as others. You've got to recognize that early and get them off the bus or out of the boat."*
>
> (Brendan Nelson, speaking to Gagel, 2019 Greatness Podcast)

It is not enough to ensure that people on teams are in the same boat. As a leader, you must also ensure that all departments, business units, and various leaders are in the same boat and aligned. One leader once said to me, *"Not only are the leaders of my department not in the same boat, but they are also beating each other over the head with each other's oars!"* That is not going to lead to success for an organization. I was once asked to come and speak to a group of about 120 engineering firm directors from throughout the world. Once we had developed psychological safety in the room, I asked them point blank, *"Are we all in one boat or 120 boats?"* Their candid answer was that they were in 120 boats and this was undermining their efforts to achieve growth and profitability. That conversation was an important first step toward building alignment within their business unit.

Figure 21.1   Gagel Model of Change Momentum.

Your ability to help people achieve change is a gift you give to those you lead that is rewarded with team and organizational success. While difficult, leading change can be rewarding. I encourage you to reflect upon the different ways people within your team and organization react to change, and to understand as a leader that different people will require different approaches to achieve change. Your ability to achieve alignment on the need for change is critical; we are all in the same boat.

Activity:

- How in tune are you with the level of comfort those you lead and work with have with change?
- How might you think about developing greater comfort with change for your team or organization?
- Are there people on your team or in your organization who are standing on the mountain? Does this need to be addressed?

Defining Organization Agility

In my dissertation research, the business units and organizations with the highest levels of agility were in the construction industry, and I am not surprised by that for two reasons: (1) the construction industry is a cyclical industry that requires our organizations to be nimble in upsizing, downsizing, and pursuing new markets to survive; and (2) every project is typically a new team of people brought together to execute with excellence, those rapidly reconfigurable resources I discussed earlier. Both require organization agility.

> "*I think one of the biggest lessons that I learned through my team which helped me be successful over a long period of time was that I had to embrace and be open-minded to change. In a competitive environment, change is constant, and that goes for everything from rules and regulations to the people that you're competing against and therefore the strengths and weaknesses that you're going up against. To be relevant in competition over a long period of time you must be open-minded and prepared to change both processes and your applications to keep the outcomes ticking along successfully over that period of time.*"
>
> (Anna Meares, speaking to Gagel, 2020 Greatness Podcast)

The term "organization agility" was first coined during the 1990s to refer to agility in the manufacturing industry as the US automotive industry grappled with declining market share (Nagel, 1991). The term "agility" was then closely associated with the software development industry after the creation of the "Agile Manifesto" in 2001 by 17 software developers who saw the need for a more iterative software design process with greater end-user involvement (Agile Alliance, n.d.). Over the last 30 years, agility has been researched more broadly across organizations of various industries (Worley, Williams, and Lawler, 2014; Meyer, 2015; Holbeche, 2023).

While I studied many of these researchers, academics, and consultants during my PhD studies, I have chosen to focus on a framework of agility documented in *The Agility Factor* by Chris Worley, Ed Williams, and Ed Lawler for this book (Worley, Williams, and Lawler, 2014). The authors used sustained financial performance as a proxy for agility and analyzed the financial data of more than 400 organizations over several decades. Of the companies studied, approximately 18% had an ROA (return on assets) above the median for their industry more than 80% of the time, and these organizations were deemed "agile." In studying these organizations, four routines of agility emerged (Worley, Williams, and Lawler, 2014):

1) **Agile Strategizing** – Agile strategizing involves developing a nimble strategy through the engagement of a significant number of stakeholders within and outside of the organization who bring diverse perspectives to the process, and then ensuring that this strategy is shared broadly, absorbed, and creates aligned day-to-day activities that move the strategy forward. The strategy is frequently revisited, and leaders are diligent in determining whether teams' and departments' actions support the organization's strategies and priorities. Agile strategizing addresses both the "identity" of an organization (our enduring purpose) and the "intent" of an organization (our strategic intent, including strategies, products, and services, which allow us to achieve our identity). Agility requires an alignment in identity, and I think of this metaphorically as a flotilla of boats.

If we are all aligned in our activities, and the boats are navigating in the same direction, then quickly changing direction becomes easier.

> *"What I think people don't understand as much is the identity piece, that there's an internal cultural conversation, how we talk about ourselves to ourselves when we're with each other; and then there are things that we say to the environment, to regulators, to market forces, to customers. We tell them who we are, and I think organizations miss the notion that sometimes these two conversations are not the same. Yes, we tell everybody that we're customer focused and when they try to reach us it's impossible to talk to somebody on the phone."*
>
> (Chris Worley, speaking to Gagel, 2020 Greatness Podcast)

2) **Agile Perceiving** – This routine encourages organizations to create flat organizations that have many touchpoints with the external environment to ensure that information about changes in the business environment is being gathered broadly and frequently. The information we perceive must then be shared (up and down the chain of command and across the potentially siloed departments of an organization) quickly and efficiently to ensure the development of informed strategies. This is about the speed and efficiency of information sharing with decision-makers.

Agile organizations also seek information from non-traditional sources to eliminate industry blinders, even when it might be difficult to convince others to take that information seriously. I remember sitting in a University of Denver meeting while I was Assistant Dean of the Daniels College of Business. We were discussing acquiring a new donor management software to track our fundraising activities. People were suggesting which universities we should talk to about this change, and at one point I said, *"Maybe we should go benchmark an organization like Google?"* Everyone paused, looked at me, and then carried on with their conversation as if I had not even spoken!

> *"I think this whole COVID-19 thing is laying bare the notion of being future-focused. Organizations were not ready. I think a lot of our talk is about being future-focused and concerned about 'this stakeholder' and 'that stakeholder' and trying to connect with them and look into the future. I think we have a lot of people who are a little embarrassed by how much the COVID-19 virus has revealed about our narrow-mindedness. We maybe talk a better game than we actually do."*
>
> (Chris Worley, speaking to Gagel, 2020 Greatness Podcast)

3) **Agile Testing** – In order to implement new strategies and tactics, we need to try new things. Interestingly, this is typically the lowest-scoring routine of the four in an organization. Agile testing requires us to adequately define the hypothesis and assign appropriate resources to the testing of new ideas. After running the test, thorough debriefing ensures that the knowledge gained during the test, even if unsuccessful, is fed into the next iteration of testing and strategies. This is often a challenge for the construction industry, as we are off and running to the next project without any formal debrief of what worked and what did not on a project. I remember working on one multi-billion-dollar project where several new Lean practices were tested. After the project was completed, I had trouble bringing together 20 people to document our learnings.

Years ago, our team conducted a gas utility industry study of technology adoption that assessed 13 technology adoptions, both successful and unsuccessful. Implementations that failed had one element in common: not assigning enough resources to adequately test the concept. How often do we give busy people more to do? The successful technology implementations over-communicated the "why" of the test and how it aligned with the overall strategy of the organization and assigned adequate resources.

> "*Testing is by far the lowest scoring of the four routines across organizations and across industries. It's not that organizations are not encouraging innovation, they are. They recognize it as fuel for the future. Most organizations are getting better at moving resources around – people, budget, time – to support innovation, that's not the one that's holding the testing routine down. The one that's holding it down is the learning part, it's putting in the time and systems to think about, 'What did we learn from the experience, that digital initiative, that failed experiment? What did we learn from that and how are we going to take that and apply it to future notions and ideas?'*"
>
> (Chris Worley, speaking to Gagel, 2020 Greatness Podcast)

4) **Agile Implementation** – Execution (doing things the organization and its people already know how to do well) differs from implementation, which is doing new things and developing new skills. I am reminded of a conversation with a global manufacturing client that was implementing new Lean practices on their construction projects. One day, I asked the leader of this program, "You are asking your people to be more collaborative with your suppliers. Have you considered upskilling them in collaboration, including how to do it and what 'good' looks like?" He had not. When medical systems began implementing doctor/patient phone conversations rather than in-person patient visits, this implementation required new skills. As a leader, it is important to think

about the new skills your team and organization need to be successful in implementing new ideas. This is also complicated by the fact that typically within an organization, multiple changes are occurring simultaneously.

> "*The thing about the change capability that gets missed a bit when people talk about this is the notion that all our change management models and processes tell you how to make one change at a time. What's happening today in VUCA worlds is we have multiple changes happening at the same time. There was a point in the book when we were trying to decide, do we call it implementing or do we call it orchestrating? I actually think that's a pretty good metaphor for the kinds of things that agile organizations are able to do. They orchestrate multiple changes so that you're working on that digital initiative, you're a little ahead of everybody else, take your foot off the gas a little bit, let us work on this human resource practice or training initiative over here so we can build up the skills as necessary, or a structural initiative takes out some layers or some intermediaries that allow the digital technology to thrive and deliver on its results. Those are multiple changes happening at once and that's what I think people are missing. They think, let's just put some more change management resources on that, and that's not going to do it. What you need is somebody orchestrating.*"
>
> (Chris Worley, speaking to Gagel, 2020 Greatness Podcast)

One key to achieving agility is an organizational and individual capacity to learn, innovate, and improvise. When our team was helping an energy company expand their construction spend from about $300 million to $1.3 billion, we were successful in implementing a continuous learning process with its primary contractors. We created a structured process for identifying challenges, generating options, and implementing solutions. Additionally, we created a system for measuring the success of these solutions and a feedback mechanism for ongoing implementation adjustments to fuel the improvement process. I believe that given the complexity of today's world, learning and curiosity must be valued by your organization, and structured processes to ensure learning, improvement, and feedback are vital.

> For my PhD dissertation research, I chose to measure agility at the business unit level rather than the organizational level because the research design was less complicated, and it seemed to make sense. That led to an "aha" moment for me during my study. Several corporations allowed me to survey multiple business units, and it was interesting to see that some business units within an organization received extremely different agility scores. For example, one large energy company divided their human resources department into two

> departments, with one focused on tactical activities like training compliance and benefits, and the other focused on the strategic side of talent acquisition and development. Which department do you think scored higher in the four agile routines? The latter, more strategic department, and that is probably appropriate. I am not certain one can measure agility at the organization level, but that is a discussion for another book!

Creating an agile organization does not happen overnight. It involves a deep understanding of your current values, organization culture, and levels of agility, followed by the focused application of resources to strengthen each of the agile routines. I have focused on the Worley, Williams, and Lawler framework on organization agility, and one other I recommend investigating is that of Linda Holbeche, which is documented in *The Agile Organization* (Holbeche, 2023). Learning about and applying the elements of organization agility form an ongoing journey to remain successful as an organization.

Activity:

- What steps might you take as a leader to understand the level of agility within your team and organization?
- How do the values and culture of your organization either support or hinder agility?
- How might you engage your team in a discussion of the need for agility?
- Which of the four routines do you believe are strengths? Which are opportunities?
- What one step might you take to improve agility in that routine?

Leader Behaviors that Support or Inhibit Organization Agility

My initial question when starting my PhD studies was, *"How can leaders create nimble organizations?"* I deeply wanted to understand the specific leader behaviors and actions that support agility because I felt this would help me coach leaders who wanted to create highly change-capable organizations. Without leadership vision and support, the journey to agility will not be smooth.

> *"If you're going to make a transformation in your organization from some sort of traditional organization to an agile organization that's probably going to take*

some level of commitment from the top. There is going to have to be somebody that says, 'This is the right thing to do, I'm going to put some resources behind this, and I'm going to put my shoulder and the shoulder of the leadership team into this and really get the organization pushed off a little bit in a direction consistent with agility.'

"The mistake that happens then is they get that started but they begin to realize that to be successful the program then needs to shift in the leadership focus from me pushing to allowing you in the middle to lead. It's the development of that shared leadership perspective. It's a very difficult double bind for an executive to say 'I want my organization to be agile and I've got to kick that off' and then hand much of the power to the middle managers to run the business so that those at the top of the organization can think about the future, play with the future, figure out these scenarios, have discussions about what we would do if something really crazy happens so when it does happen we're ready to go. I think there's a leadership shift that must take place there and it can be threatening. 'I rose to the top, I got to the top of the organization, I have the power now, I should use it.' To start agility, and to keep agility going, I must give that power away."

(Chris Worley, speaking to Gagel, 2020 Greatness Podcast)

My PhD research study identified four key categories of leader behaviors that lead to the presence of agile routines in organizations (Table 21.1):

Visionary Leadership Behaviors: When we think about the leadership theories and frameworks discussed throughout this book, the ability to create a vision that resonates with the team and organization's members is often defined as a critical leadership skill. Thinking back to John Kotter's SAM (set direction, align, and

Table 21.1   Gagel Research: Leadership Behaviors that Support Organization Agility.

Behavior Category	Description of Leadership Behaviors
Visionary	Leadership behaviors that create a clear organization purpose and mission that define the "why" of the organization's existence
Exploratory	Leadership behaviors that support a culture of discovering new ways to solve problems and conduct business
Latitude	Leadership behaviors that provide employees with a high degree of freedom and responsibility in achieving work results and resolving issues
Reflective	Leadership behaviors that cause the leader to challenge their own assumptions and create mechanisms for the organization to do so as well

motivate), people want to have a clear picture of the finish line (Kotter, 2012). What is the purpose of our organization? Our team? What is the difference we are trying to make in the world? What would not happen if we did not exist? The answers to these questions must be brought to life by the organization's leaders.

Vision is important for agility because it is about alignment. If we are all aligned on the vision of why we exist and the values we believe are most important, it is a great deal easier to pivot on the "how" we do it, our strategies for success, and that creates greater agility. As a leader, it is important to regularly test people's understanding of the vision. I go back to the story I shared of the CEO who felt he was adequately articulating the vision of the organization, only to find that his leadership did not agree. As leaders, we often believe we are overcommunicating the vision, probably because it is clear in our minds. Testing whether others understand the vision is critical.

Exploratory Leadership Behaviors: The important phrase is that these leadership behaviors "support a culture of…" exploration, that the embedded behaviors and assumptions of the team and organization affirm the need to continually consider new ways of working. This calls for us to break out of the paradigms we have about how work is currently being done, which is difficult for many people, teams, and organizations.

When creating and delivering a post-graduate course on design thinking at the Australian National University, I created a student exercise that exemplified how difficult it is to break free from our existing mental models. I broke the class into three groups and assigned each an every-day item within a specific context, then I explained to them my personal frustration with that item: (1) a whiteboard marker in a classroom. These are frequently dried up, and I discovered that in Australia alone, we throw away about 3 million per year; (2) a to-go coffee cup and lid at the university café. Although I do not drink coffee, I often enjoy a hot chocolate before my lecture, and I can never quite get the lid on right, causing a constant fear of "wearing" my hot chocolate to my lecture; and (3) a water bottle at my cross-fit gym. As I move from station to station, I am always forgetting it, knocking it over, and sometimes even leaving it at the gym.

I asked the students to use the Stanford University d.school framework of design thinking to create a new solution for these situations (Hasso Plattner Institute of Design at Stanford, n.d.). Invariably, at least one group would come back having created an iterative improvement to the current solution, such as a whiteboard marker that is refillable, a refillable coffee cup with a more secure lid, or a gym bottle with a microchip for tracking. Some groups grasp the point, that I'm asking them to go back to the need we are solving for

> and break free from the paradigm of the current solution. It is not about a whiteboard marker; it is about the need to be able to convey a visual idea, and maybe that involves a magic board where you use your finger to draw. The need at the gym is to stay hydrated, and might not involve a water bottle at all, but instead tubes coming out of the ceiling, and you have a unique attachment around your wrist that allows you to access water. Breaking free from our pre-existing ideas about how work is completed can be just as difficult.

Think about how this applies to your team or organization. What process, solution, or model are you stuck in? When was the last time you held a "crazy idea" session to brainstorm new ways of doing business? Edgar Schein's definition of leadership is to find things that need to be better, that need to change, and making it happen (Schein and Schein, 2018). This is the essence of exploratory leadership behaviors.

Latitude Leadership Behaviors: Latitude leadership behaviors provide team members with the freedom to determine how best to accomplish their assigned roles and responsibilities. What was surprising about this result was that the research indicated that leaders who leaned toward an almost "laissez-faire" leadership style (nearly to the point of "not here when I need you") led teams with high levels of agility. Upon reflection, this makes sense. If you work for a leader who is not always available, you are forced to be agile in solving problems and accomplishing work, which creates more agility in the team. I believe this is one of the reasons that construction companies scored so highly in my research study. People out on construction jobsites are typically given a great deal of latitude in how they accomplish work and solve problems.

> *"You're more likely to invest in training and development for your people so that they're empowered to act. You're more likely to encourage information sharing and have systems and processes that support that. You're more likely to be able to move swiftly as the market changes. And then you're more likely to be able to be flexible."*
>
> (Linda Holbeche, speaking to Gagel, 2019 Greatness Podcast; Holbeche, 2023)

Worth noting is that different people require different levels of latitude, and profiles such as the DiSC profile can help you understand these differences. For example, people with a high level of "C" feel there is a "correct" way of doing things, and most of the time, they are the ones who define what "correct" looks like. This can leave little latitude for other ways of doing things. Some people like to be told exactly what to do. Others (like me) prefer to have a goal set and then determine how to achieve it. I was interviewing for a role once and a couple of the people I interviewed with asked whether I would feel comfortable working in a role where much of my game plan was defined for me. I thought to myself, "That is okay if I can add my ideas to those defined for me," and I took the job. Wow, was I wrong! This leader had the only good ideas, and I was given no latitude in how I executed my leadership role. It is important as a leader to adjust your leadership behaviors to fit the needs of each person on your team.

Reflective Leadership Behaviors: When I think of reflection, I think of being deeply present to an idea or thought. I think of having the courage to fully unpack that idea or thought, to challenge my assumptions and mental models of how something "is" or how it should be. The ability to reflect as an individual or team is critical and requires time. This might start with a leader saying, "Hey, let's sit down for an hour and think about how process XYZ is working and how we might improve it," with an agenda only to deeply reflect upon the process and how it is working or not working; or perhaps to challenge the fact that the process exists at all; or perhaps to brainstorm what an entirely new process might look like. To create this ability to reflect, people must be open to new ideas; they must be focused on the task at hand and not 100 other things on their to-do list; there must be enough psychological safety present for people to share "crazy ideas"; and leaders must not privately condone behaviors that promote creative thinking.

My research study also identified leadership behaviors that inhibit organization agility (Gagel, 2021):

- **Being Overly Structured** – Structure within a team is critical, but if a team is overly structured, people will not collaborate or bring outside thinking into problem solving, which can in turn reduce organization agility.
- **High Levels of Power** – Power is not a dirty word! But when a leader exhibits too much power, there can be a corresponding reduction in psychological safety, which in turn inhibits "crazy ideas" and organization agility.

I believe the ability of a leader to create an organizational culture where people understand that change is necessary and ever-present is critical in today's complex and continually-evolving business landscape. It is up to each leader to determine how much agility is required of your teams and organization, and then behave in a way that supports agility. Your actions as a leader have a tremendous impact on the agility of your team and organization.

Activity:

- Take some time to reflect upon the leadership behaviors that support organization agility. Where do you believe you excel? Where might you improve?
- Do you believe you exhibit leadership behaviors that might suppress organization agility? What might you do differently?
- How might you help those you lead exhibit more agility?

The Role of Organization Ambidexterity in Organization Agility

When I think back to the example shared earlier of the private-equity-owned firm that was struggling to achieve its growth objectives, the concept of organization ambidexterity was critical to solving the problem. The firm was not making deliberate decisions about the allocation of resources to existing business models versus new potential markets. Typically, existing business models win and gain more of an organization's resources, while new business model initiatives might struggle along with limited resources.

James Gardner March provided much of the early thinking on organization ambidexterity, which focuses on making "deliberate" decisions about the allocation of resources to **exploiting** current strategies and business models and **exploring** future strategies and business markets (March, 1991). I say "deliberate" because I believe many organizations make these types of decisions in a rather ad hoc manner. An ambidextrous organization is thinking long-term and realizes that the golden goose business model of today may be the shriveled-up prune of tomorrow. Think Kodak or Blockbuster. These organizations were not thinking in an ambidextrous way.

During my time at one organization, decisions were driven by entrepreneurial individuals who seized an opportunity to build a business around a passion that fit within the purpose of the organization, but limited resources were allocated to their efforts. In fact, I remember being told, "Sure, you can start this business. We don't really have any money for you to market it or hire people, but good luck!" It worked out, but it took some time for our leaders to understand the value of being ambidextrous as an organization by allocating resources to new ventures.

Ambidexterity is also a concept that applies to departments, teams, and individuals. Let's say you are an estimating department. Should you be deliberately applying resources to exploring the application of AI (artificial intelligence) to your estimating process, and if so, how much money and time should you devote to that effort? How many people? If you are a project team, should you perhaps be carving off resources to try a new way of tracking subcontractor productivity? Teams and departments are constantly making decisions about the allocation of resources. Organizational ambidexterity brings discipline to the process of considering resources for existing strategies and tactics versus emerging strategies and tactics.

Individual ambidexterity involves successfully utilizing your current skillsets while exploring new opportunities to thrive. As a management consultant, I was unable to volunteer for nonprofit boards, given my extensive travel, and I missed giving back to my community. I was also becoming weary of the travel and a culture that did not value leadership skills. I deliberately carved off some personal resources—time and money—to go back to school and complete a Master's in Nonprofit Management at Regis University. I was not entirely sure what I was going to do with that degree, but it allowed me to explore new possible business models for me, and how I could continue to serve my life passions and earn a living. This eventually led me to leave management consulting to become CEO/President of The Women's Foundation of Colorado.

Ambidexterity at the organization, team, and individual levels is an important concept to understand as a leader. Devoting too many resources to exploiting current business models and skills could result in obsolescence, and devoting too many resources to exploring future business models and skills could result in over-extension and failure. Finding the right balance and making deliberate decisions about the allocation of resources to exploitation versus exploration is key.

Activity:

- How well does your team or organization make deliberate decisions about the allocation of resources to existing versus emerging strategies and tactics?
- How might you, as a leader, improve the ambidexterity of your team or organization?
- What does ambidexterity look like to you as an individual?

A Note on Scenario Planning

During my PhD studies, I had the incredible opportunity to take a course on scenario planning with global expert Dr. Thomas Chermack (2011). Beyond suggesting you read his insightful books on the topic, I would like to share a few observations about scenario planning. I believe this is one of the most beneficial exercises your organization can embark upon to help equip you for an uncertain future business environment.

- Traditional strategic planning assumes the organization is at Point A and wants to achieve Point B. However, scenario planning assumes there are multiple possible futures at Point B, and that strategic planning is not a straight-line process.
- Scenario planning is not about predicting the future business environment. Rather, it is about making business decisions that meet the needs of the most possible future business environments.

- The scenario planning process is most effective when people from different layers and departments are involved instead of pulling together the executive or senior leadership team. This includes people who are closest to the customer and the work, such as superintendents and project engineers.
- Key to effective scenario planning is the identification of the elements of your organization's business environment that are: (1) most difficult to predict; and (2) have the greatest potential impact on the business. Examples include inflation, labor shortages, demographic trends, and political changes.
- Scenario planning typically does not focus up "black swan" events like pandemics.
- Gathering information from non-traditional sources is critical to the process. For example, if you are a gas utility engineering leader and all your information is gathered by reading industry magazines and attending industry conferences, you are NOT gathering information from diverse sources.
- Again, scenario planning is not about predicting the future. Most strategic planning processes assume you are starting at Point A and trying to get to Point B. Scenario planning assumes you are at Point A and have no idea what Point B is, and that there might be a Point C and Point D and Point E, given the possible changes in the business environment. The key is to make business decisions that mitigate risk and maximize reward in as many of those "points," or business environments, as possible.

Just a taste of the concept of scenario planning!

Conclusion

One of my larger clients has shifted from bidding all their large construction programs to at times sole-sourcing multi-billion-dollar programs. Why? Speed to market. The business environment is changing so quickly that if this company takes the time to gather bids for the project, the market will have shifted out from underneath them. This is not an isolated instance. Another large owner client in the beverage industry almost entirely missed an emerging market because they could not build manufacturing capacity quickly enough.

"One of the things that I learned a long time ago is how powerful assessment is. If your organization has been around for more than five years you've beat the odds. If you're a 10, 20, 50, 100-year-old organization you've really beat the odds. If you're surviving, then there's something about your organization that is contributing to

> *that, and I think it's one of the four routines. If you want to become more agile that means understanding which of the four routines are strong and weak inside your organization and then beginning to target the weaker ones and raising those up without losing the others. Maybe you're good at strategizing, hold on to that one and add a routine to your quiver so that you've got something else to rely on."*
>
> (Chris Worley, speaking to Gagel, 2020 Greatness Podcast)

The rate of change is increasing in the world, which in turn drives the need to employ organization agility and ambidexterity at the organization, business unit, team, and individual level to remain competitive. Those organizations that do so will survive, and those that do not… well, we know how that story ends. As a leader, you have the power to equip your people with the tools they need to embrace and execute change, and I encourage you to make this a priority.

References

Agile Alliance. (n.d.). *What is the agile manifesto?* Available at: https://www.agilealliance.org/agile101/the-agile-manifesto/ [Accessed: 16 June 2024].

Belasco, J. A. (1991). *Teaching the elephant to dance: The manager's guide to empowering change.* New York: Plume.

Chermack, T. J. (2011). *Scenario planning in organizations: How to create, use, and assess scenarios.* Oakland: Berrett-Koehler Publishers.

Conner, D. (2012). 'The real story of the burning platform,' Reply-MC.com, 2 September. Available at: http://www.reply-mc.com/2012/09/02/the-real-story-of-the-burning-platform/ [Accessed: 12 July 2024].

Gagel, G. (2021). 'The effects of leadership behaviors on organization agility: A quantitative study of 126 U.S.-based business units,' *Management and Organizational Studies*, 7(1). Available at: https://www.researchgate.net/publication/349091345_The_Effects_of_Leadership_Behaviors_on_Organization_Agility_A_Quantitative_Study_of_126_US-Based_Business_Units [Accessed: 16 June 2024]

Gagel, G. and Holbeche, L. (2019). *Linda Holbeche discusses her book The Agile Organization* [Podcast]. 10 June. Available at: https://open.spotify.com/episode/5hF8JOEOH0kIQ1RGrZdMZS [Accessed: 28 April 2024].

Gagel, G. and Meares, A. (2020). *Anna Meares discusses leadership and team lessons from being an Olympic athlete* [Podcast]. 17 February. Available at: https://open.spotify.com/episode/3k1RqB8K3aiLVG5lRgkRol [Accessed: 28 April 2024].

Gagel, G. and Nelson, B. (2019). *Dr. Brendan Nelson discusses leadership and personal values* [Podcast]. 10 May. Available at: https://open.spotify.com/episode/7cadZRiknfvUF63YeWcAlj [Accessed: 28 April 2024].

Gagel, G. and Ransom, H. (2022). *Holly Ransom discusses her book The Leading Edge* [Podcast]. 30 September. Available at: https://open.spotify.com/episode/0M5G6yRwqvbBlKa302wmKb [Accessed: 1 May 2024].

Gagel, G. and Worley, C. (2020). *Dr. Gretchen Gagel interviewing Dr. Christopher Worley on organizational agility*. 5 October. Available at: https://www.youtube.com/watch?v=i0NqMs5tkPk [Accessed: 16 June 2024].

Hasso Platner Institute of Design at Stanford. (n.d.). *An introduction to design thinking: Process guide*. Available at: https://web.stanford.edu/~mshanks/MichaelShanks/files/509554.pdf [Accessed: 16 June 2024].

Holbeche, L. (2023). *The agile organization: How to build an engaged, innovative and resilient business*. 3rd edn. London: Kogan Page.

Kotter, J. P. (2012). *Leading change*. Boston: Harvard Business School Press.

Lewin, K. (1947). 'Frontiers in group dynamics: Concept, method and reality in social science; social equilibria and social change,' *Human Relations*, 1(1), 5–41. Available at: https://doi.org/10.1177/001872674700100103.

March, J. G. (1991). 'Exploration and exploitation in organizational learning,' *Organization Science*, 2(1), 71–87. Available at: https://psycnet.apa.org/doi/10.1287/orsc.2.1.71.

McChrystal, S., *et al.* (2015). *Team of teams: New rules of engagement for a complex world*. London: Portfolio.

Meyer, P. (2015). *The agility shift: Creating agile and effective leaders, teams, and organizations*. New York: Routledge.

Nagel, R. N. (1991). *21st century manufacturing enterprise strategy: An industry-led view*. Available at: https://www.researchgate.net/publication/235112061_21ST_Century_Manufacturing_Enterprise_Strategy_Report [Accessed: 16 June 2024].

Schein, E. H. and Schein, P. A. (2018). *Humble leadership: The power of relationships, openness, and trust*. Oakland: Berrett-Koehler Publishers.

Worley, C. G., Williams, T., and Lawler, E. E., III, (2014). *The agility factor: Building adaptable organizations for superior performance*. San Francisco: Jossey-Bass.

Part V

"Us" – Leading and Influencing Teams and Organizations – Conclusion

Developing strong skills as a leader of teams and organizations not only fuels your individual success but also enables the success of others. Teams and organizations are complex systems filled with people who are at times difficult to predict. I believe that leading teams and organizations is one of the hardest things I have ever done, and one of the most gratifying. There are few things in life more satisfying than helping a team or organization define its purpose, set goals and objectives, develop strategies and tactics to achieve these aims, and then celebrate success. Just as gratifying is developing the change muscle that enables these teams and organizations to embrace the agility that is essential to long-term success.

I cannot overstate the importance of values and culture in effectively leading teams and organizations. How you as a leader not only lead from your own personal values but also clearly define the non-negotiable values of your team and organization sets the tone for behaviors and expectations. A team or organization's culture, while at times difficult to describe, is a strong determinant of how success will be achieved. Being present to and curating the desired culture is one of the most important things you will do as a leader.

Finally, it truly is about the people. I hire the brightest I can find and empower them to use their brains in collaboration with others to drive success. I invest in people's development and root for their success. I believe that success as a leader is measured by the success of those around you.

Is all of this messy? Of course. Is leading teams and organizations hard? Of course. But you have the capabilities if you begin with the humility that you do not have all of the answers, and work to build the skills discussed in this portion of the book. I believe in you!

Building Women Leaders: A Blueprint for Women Thriving in Construction, First Edition. Gretchen Gagel.
© 2025 John Wiley & Sons, Inc. Published 2025 by John Wiley & Sons, Inc.

Part VI

"Us" – Leading an Inclusive Construction Industry

22

Leading an Inclusive Construction Industry

Introduction

I was standing at an APGA cocktail party in Sydney in December 2023 after wrapping up Cohort 7 of APGA's Women's Leadership Development Program. One of the APGA Board members, Scott Pearce (CEO of CNC Project Management) was addressing the room. Scott mentioned how important he felt the women's leadership program was to the industry and how attending dinner with the group provided him with more empathy about being in the minority in a room. I had a "light bulb moment." Each woman who attends this program is not only receiving the gift of leadership and personal development; they are also giving a gift to the pipeline industry, in that they are helping this industry become more inclusive just by showing up.

I believe we have an ability and a responsibility to lead not only ourselves, our teams, and our organizations, but also to help shape the future of the construction industry. I think of my friend Jan Tuchman, retired Editor-in-Chief of ENR, and the multitude of awards Jan has received in recognition of all she has done to lead the industry. I think of my friend and mentor Hugh Rice, Chairman Emeritus of FMI, who leads not just via his significant roles in the construction industry but also by being a role model of inclusive, respectful leadership. As leaders, we have the power to shape the world around us—our families, our communities, our teams, our companies, and even our industries. Leading the industry comes from the heart and from a genuine place of caring about the construction industry.

> *"Male allies have the power to lead change and drive inclusion in the construction industry. I think of the male directors of the Australian Constructors Association who came together in 2019 to create a requirement for every male Board member to name a female co-director. This bold move changed the face of construction*

Building Women Leaders: A Blueprint for Women Thriving in Construction,
First Edition. Gretchen Gagel.
© 2025 John Wiley & Sons, Inc. Published 2025 by John Wiley & Sons, Inc.

> *industry leadership in Australia, and while the first women to join the Board described it as initially awkward, a new norm was developed within a year, and it was as if they had always been there. This is leading to courageous change in our industry."*
>
> (Redwin, 2021)

In this chapter, I encourage you to reflect upon those you admire for leading in the construction industry. Consider your own efforts to help shape the future of our industry, and the ideas from this chapter that might enhance those efforts. How might you continue contributing to the success of our industry? What actions might you take, and how might you inspire those around you to take action as well?

Individual Actions to Lead the Industry

The actions we take as individuals shape the future of the construction industry, and I believe leading the construction industry begins with these steps:

1) **Love the Industry**
   I love the construction industry. When people ask me why I am so dedicated to the construction industry and the people in it, my response is clear: "Do you want to drive on a road? Eat lunch in a lovely restaurant? Have your baby in a hospital? These things are all constructed!" As I have mentioned throughout this book, I love the construction industry because we humbly build and maintain the assets of civilization.

> *"I think our industry is a great industry. I love what we do, I love building, I love the people. I get up every day just excited to be in this industry and I feel it's an honor to be able to point at something and say, 'Yes, we were part of that.'"*
>
> (Denise Burgess, speaking to Gagel, 2023 Greatness Podcast)

In *The 7 Habits of Highly Effective People,* Stephen Covey talks about two circles: the larger circle of everything we care about, and the smaller inside circle of everything we influence and control (Covey, 2020). Covey makes the point that if you focus on the things you care about that you do not control, you will experience tremendous frustration. Instead, Covey encourages us to focus on what we control, and in doing so, the circle of what we influence will expand.

One of the ways we expand that circle of influence is by sharing our own stories of success. In Australia, there is this cultural element called "tall poppy syndrome," meaning people should not boast or stand above their "mates." I agree, boasting is egotistical. Yet sharing our success stories as women in the construction industry (in a gracious way that acknowledges the many people on our teams that are contributing to our success) is important. Stepping up to lead is important. Be a role model, toot your horn. You belong.

Each of you, with each action you take, expands your circle of influence and makes your mark on the industry. Share your love of our industry—your pride in what we do—with others outside of the industry. Be the beacon that is an inspiration to others. Embrace our industry for the wonderful things we accomplish and the tremendous people who show up each day to make the world a better place through our work.

Activity:

- What is your answer when someone asks you why you work in construction?
- In what ways do you demonstrate your love for our industry? How can you share that love with others?
- What actions might you take to focus more on what you control and less on what you do not control in our industry? How might you expand your circle of influence?

---

2) **Take Individual Action**

I recently had the pleasure of having coffee with a new friend, Georgie Wrighti (General Manager of Jemena), to discuss how Jemena could continue supporting the 30 or so women who have completed the APGA Women's Leadership Development Program. During our coffee, Georgie told me the story of starting as an engineer in her first drywall plant, where she and another woman had to travel a significant distance in the office to use the restroom. Georgie, at the age of 23, took it upon herself to design and build a women's restroom in the plant. That is taking action!

We all need to pick our battles, and I do not walk around the construction industry with a chip on my shoulder over every slight. However, I firmly believe our individual actions make a difference in our industry. I am positive that politely and repeatedly asking if there is a women's portable toilet on the construction sites I visit makes a difference. One of the speakers during the 2024 NAWIC Victoria luncheon put it best: "I'm tired of standing before you talking about toilets!" This is a fixable problem in our industry, and we can make it a priority.

What is good for women is good for everyone. I am reminded of a story told by a retired female mining executive about her first experience as a material testing engineer in a mine. The walls above the material testing benches were covered in pictures of naked women. This woman started "spilling" a bit of coffee each morning on the wall ("oopsy") and taking that picture down until her area was picture-free. A man approached her one day in the mine, asking if she was the one taking down the pictures. Her initial thought was, "I am in trouble now." But she confidently said, "Yes." The man thanked her profusely, saying he was deeply religious and found the pictures offensive, but did not feel he could say anything because of the ridicule he would face from his peers. She emphasized the point that what is good for women is about having respect at work, and that is good for everyone.

Each action you take helps lift the bar of expectations and improves what is already a great industry. You can make a difference in our industry with your individual actions to be inclusive and help shape a respectful industry where everyone feels they belong.

Activity:

- Think of a time when you or someone else took an individual action that demonstrated leadership in our industry. How did that make you feel?
- Perhaps there has been an instance where you felt powerless to take action. What might you have done differently?
- What other individual actions might you take to make our industry more inclusive?

---

3) **Raise Your Hand**

Our Executive Leadership Team at FMI went through a company revisioning process in 2003, given it was our 50th anniversary. Our team met in a different city every three-to-four-weeks for nearly a year, reevaluating our purpose, vision, mission, values, business units, and strategies, and it was gratifying to play a role in helping the organization define success for the next 50 years. As this process ended, we decided to name a new President. One of my male colleagues advocated for me to raise my hand, but I did not. I did not believe the company was ready for a 40-year-old female President, and I did not think I had what it took. Looking back, I realize I probably did have what it took and perhaps should have raised my hand. I learned a valuable lesson.

As Brandy McCombs, President of IBC and first female Chair of The Builders, a chapter of AGC of Kansas City, shares, *"I'm going to give advice to everybody listening. Don't wait, because I waited. I was asked to be the chair and secretly deep down inside I wanted to be the chair and you would think out of all the people in this world that could ask for the seat at the table it would be me. But I did not and I'm really glad that someone else saw it in me to do it. It was almost like a blessing in disguise."*

(Brandy McCombs, speaking to Gagel, 2023 Greatness Podcast)

The more senior the position you hold, the more influence you have within your organization—and in turn, the industry. Raising your hand, asking for that next promotion, that next leadership position, is another important way you can influence our industry. We need to see more women in senior leadership positions in the construction industry.

About halfway through an MBA school lecture at the Australian National University one night, a student was looking a bit dejected. I asked what was bothering her, and she responded by saying, "There are so many bad leaders in the world." I have to agree. I looked at her and said, "The opportunity is for you to be an incredible leader, for you to be a shining example of what great leadership looks like, and for you to lead, develop, and mentor hundreds of people who themselves become great leaders." By raising our hands and stepping up as great leaders, we inspire others.

Some of us may not be in a position to take on additional responsibility, and that is okay too. I stepped away from my leadership position at FMI because my kids were seven- and eight-years-old and I was missing too many hockey games, too many school functions, too much of their time growing up. When my mother was in the last stages of Alzheimer's, I once again stepped down from my position as Assistant Dean at the University of Denver to spend more time with her, and I credit a close friend for giving me the courage to do this, as my time with my mom was an incredible gift. It is okay to not raise your hand when circumstances prevent you from doing so.

As Annabel Crookes, a Director of Laing O'Rourke and first female President of the Australian Constructors Association (ACA), shares, *"When I was first asked to nominate for President I said no for a year because I wasn't certain I had the skills to succeed, or the energy given my role at Laing O'Rourke and my family responsibilities. I finally said yes because I wanted to help shape the future of the industry and show that women with different styles can succeed; and because Laing O'Rourke provided additional support to ensure my success"* (Crookes, 2024).

Opportunities exist for each of us to raise our hands. I encourage each of you to step into whatever role is right for you at whatever time in your life. Believe in yourself and have the courage to go for a position you might believe is a stretch. You can do it!

Activity:

- What opportunities might you raise your hand for? What might be stopping you?
- How might you leverage your support system to gain the courage to raise your hand?

---

4) **Join Associations**

When I started the Owner Services Group at FMI in the late 1990's, I had no choice but to become involved in industry associations, as this was the best way to network and build brand recognition for our group. I remember the early days of the Construction User Roundtable (CURT) and the Construction Owners Association of America (COAA). I dove into the Construction Management Association of America (CMAA) and the Construction Industry Institute (CII), and the friends I made at each of these associations remain friends today. Yes, back in the "dark ages," was it challenging to be one of the few women at these meetings? Of course. But I looked at it as an advantage. How many other tall, blonde "Gretchens" were there walking around?! People remembered me, and they respected me and my expertise in the industry.

"Raise your hand when there's opportunity to raise your hand, get involved when there's an opportunity to get involved, join a committee, get a membership, get the cheapest membership. But then start to get involved. Don't put your toe in, just go in the deep end of the pool and you'd be surprised at how many people will help you and support you and how much you'll just enjoy it."

(Denise Burgess, speaking to Gagel, 2023 Greatness Podcast)

Many of you might be thinking, "I'm so busy as it is, how do I fit in one more thing?" You can start small by joining your local AGC (Associated General Contractors) or ABC (Associated Builders and Contractors) chapter in the US, or your local National Association of Women in Construction (NAWIC) chapter in the US or Australia. Many industries have their own associations—such as energy, mining, and transportation, for example. Even if you only show up at one event a year, it makes a difference because you are saying, "I care about my industry and the people in it," and you are helping us build a more diverse industry. By sitting on the AGC's national Diversity and Inclusion Committee, I make a difference.

Joining associations in our industry is an important way to lead, help shape the future of the industry, and network. I encourage you to give thought to this leadership strategy.

Activity:

- What associations might you join?
- If you already belong to an association, is there a way you might become more involved by attending an event or joining a committee?
- Perhaps you might consider leading an association?

---

5) **Join Groups Working to Change the Industry**

Just prior to my move to Australia in 2018, our organization worked with Stephen Mulva, then-Director of CII, to help create a *Manifesto for Change* for the construction industry. Stephen's team and the Continuum Advisory Group team brought together industry leaders in three focus groups in the US to define the industry's issues and create a proclamation of the need to address them. Advancing this type of change is difficult and can be threatening to others who enjoy the status quo, but disruption is necessary to move an industry forward. Years earlier, CII led the charge to standardize industry safety measures and require contractors to achieve certain safety levels to compete for projects—another great example of CII driving change in the industry.

When I arrived in Australia, I sought out people doing similar work. Jon Davies, then-newly appointed as the CEO of the Australian Constructors Association (ACA), suggested I meet with Gabrielle Trainor, AO, Chair of the Construction Industry Culture Taskforce (CICT) for Australia, and in January 2021, I joined the Steering Committee (Construction Industry Culture Taskforce, n.d.). This group, with funding from the New South Wales and Victorian governments, has taken a research-based approach to improving the culture of the construction industry by promoting the adoption of a Culture Standard, and the progress has been amazing.

"If you look at the statistics here in Australia it's something like 12% gender diversity on a good day in the construction industry. It's an industry with very high rates of marriage breakdown and mental illness, high suicide rates, and all of this contributes to an industry where you're really shutting out half the workforce. We set out to try and work out the best way to really move the needle, what mechanism

could we use between all of us to actually change the fundamentals and really make a difference to the industry. We developed a Culture Standard which would become part of the procurement process. Contractors bidding for these big projects would have to pay attention to the various elements that we identified. They would have to pay attention to the culture of our industry, to improve it, to make it a much better place for women and diverse groups more generally to belong, to pay attention to things like mental health and wellbeing, to pay attention to amenity and facilities for women, to pay attention to pay gaps, to pay attention to all the things that contribute to it not becoming a welcoming and employer of choice workplace, particularly for women."

(Gabrielle Trainor, speaking to Gagel, 2021 Greatness Podcast)

There are many more examples I could share of groups that are working to effect change in the construction industry. These groups create power by simultaneously lifting the voices of multiple stakeholder groups, and you joining these groups lends your voice to their efforts as a leader in the industry.

Activity:

- What organizations do you feel are working to build a more inclusive construction industry?
- How might you add your efforts to theirs?

6) **Start a Group or Create an Event to Work on Changing the Industry**

While I encourage people to first consider joining an existing group, at times there is a need to create a new group or event to lead the efforts of creating a more inclusive industry. One of the outcomes of the APGA women's leadership program was the creation of the Women in Pipelines Forum, initially Chaired by industry leader Cara Robb. *"I know that I wasn't alone with that feeling that we want to network, we want to connect with other women in our industry, and we need a group. I think over a Greek dinner quite a few of us brainstormed about the idea, and Steve approached me before the 2021 virtual convention and said, 'We'd really like you to lead it as chair.' We just want to make sure that there's a space to promote women in the industry, to advocate for the industry, and to try and get more women into the industry as well"* (Steve Davies and Cara Robb, speaking to Gagel, 2024 Greatness Podcast).

> "When I joined our leadership team at Cerris (formerly MMC Corp) one of the first things that I did was do an analysis of the women in our organization because many times there might be only one female in the entire organization outside of an administrative role and women didn't have much of a network in their own space. I proposed the idea of creating a women's professional network for our organization and luckily it was met with support from our male ally leaders.
>
> "Then I attended an event at a local college to prepare female athletes for professional life. I thought, how great would it be if we could create something like this for the architecture, engineering, and construction community. As I had witnessed in our own organization, there were probably many women out there that might only have one or two other women in their entire organization and many times none of them have a female leader to have as a mentor or as an ally.
>
> "I went to several other powerhouse women in our community and pitched the idea of creating a program focused on women in the built environment and we called it On the Rise. It's a half-day program with panels, guided roundtable discussions with female leaders at your table, and a keynote speaker. We started it in 2020 with about 150 people, did it all on our own time, found sponsors to help fund it and had support from our own companies to help fund it. We've had one ever since and it's grown to a little over two hundred people and several sponsors that are still very much supporting it. I feel lucky to have as much support, especially in our community, for this endeavor."
>
> (Erica Jones, speaking to Gagel, 2024 Greatness Podcast)

As Paula Gerber, Law Professor and Associate Dean (Education) at Monash University and Founder of the National Association of Women in Construction (NAWIC) Australia, shares, "*I spent 10 years overseas, five years in London and five years in California, and when I came back to Australia there were two things that I was very passionate about seeing created in Australia. One was a Master's in Construction Law like the one I did at King's College in London because it was the only one in the world at that time; and the other was to join NAWIC and I was shocked to find that it didn't exist and that there was no association for women in construction. I had been a member of NAWIC in Los Angeles, in the Santa Monica chapter, and I absolutely loved it because the firm that I worked with in California was a boutique construction law firm and every lawyer in the practice was male. I was craving a connection with the more diverse part of the construction industry practice and I stumbled upon NAWIC. That's where I found my people and I really enjoyed getting to know all these amazing, dynamic, strong successful women. I came back to Australia after 10 years away and I really didn't have a network here. I thought, well, NAWIC is a place where I met people in California, I'll join it here, but it didn't exist. I started calling every woman who ever picked up a hammer and I found just incredible enthusiasm for having an organization for women in construction. I think if I'd known how big it was going to get and how much time it was going to take, I probably wouldn't have embarked on it, but ignorance can be bliss and I just started on this journey, and it took off like wildfire."* (Paula Gerber, speaking to Gagel, 2022 Greatness Podcast).

All of you do not need to run out and start a new group or event! But I do applaud those who see a need and have the courage to create a group or event to fill it.

> **Activity:**
>
> - Is there a need for a new organization or event in your built environment community to help fuel progress for women and other underrepresented communities?
> - What might you do to explore this opportunity?

7) **Find Like-Minded People**

Back when I worked at FMI, I printed a large sign that said, "WILD DUCKS CONGREGATE HERE" and tacked it to the bulletin board above my desk (Figure 22.1). I do not remember what specific incident prompted me to make this sign, but I do remember being frustrated by a construction industry that at times was resistant to change. People who shared my frustrations (other "wild ducks") were attracted to that sign, which led to many deep conversations about how to impact the industry. Miraculously, I stumbled across this sign again, faded and a bit torn, while writing this book!

I am a collaborative person by nature, and one of my greatest joys in my career has been the fun of finding like-minded people who also want to make a difference in the industry. I think of Jon Davies, CEO of the Australian Constructors Association, for not only his individual efforts to disrupt the construction industry but also for actions like selecting me as the first female chair of the Judging Panel for the Australian Construction Achievement Award. Jon understands the importance of placing women in visible roles in our industry. He is a like-minded individual—a "wild duck."

I encourage you to be a wild duck, to find other wild ducks out there with the courage to step up and lead change in our industry. They need your help, and it's fun.

Figure 22.1   WILD DUCKS CONGREGATE HERE Sign Circa 1996.

Activity:
• Who are the other "wild ducks" out there that are driving change in our industry? • How might you engage with them? Lend energy to their efforts? Collaborate?

8) **Engage in Social Media**

On March 14, 2021, I received a message on LinkedIn from Kevin Sears, Vice President at JE Dunn Construction in their Kansas City headquarters.

> "Dr. Gagel, I just wanted to say thank you for connecting and that I really enjoyed your book. I came across your TED Talk on LinkedIn and learned of your connection to both construction and Kansas City. It was the right read for Women in Construction Week! I want to be the most empathetic and supportive supervisor I can be for the women on my team. Your book and my discussion of it with my wife helps to reinforce my mindset and leadership choices. Sincerely, Kevin Sears"

I immediately reached out to Kevin to ask him if I could anonymously repost his message on LinkedIn, and Kevin said, "Feel free to use my name!" Kevin is a person who understands the value of putting a statement out there that helps shape the future of our industry.

I am not advocating for each of you to become a social media sensation! But just paying attention to what others are saying or doing on social media can make a big difference. I could tell from the LinkedIn posts from the 2023 APGA conference that the number of women presenting had increased significantly (from five women presenters in 2018 to 16 women presenters in 2023), and the energy through LinkedIn was infectious! Consider how you might engage in the conversation about the future of our industry.

I know when one of my LinkedIn posts has struck a nerve when it receives 4,000 views in 48 hours, as this post from November 2023 did:

> "Do you know what the number one micro-aggression against women is in the workplace? Being talked over. Believe me, I have experienced and witnessed it all too often.
>
> So what can we all do about this? How do we actively advocate for underrepresented people at work and make them feel included? Many of us understand that inclusion is important, but how do we use our privilege, be that the color of our skin or our leadership position, to advocate for the under-represented?
>
> My friend Jeffery Tobias Halter, Gender Strategist joined the Greatness podcast to discuss "active male engagement in advancing women" and advocacy for all underrepresented groups. As Jeffery states, men are 75% of leadership in most companies and are 75% of the solution in advancing women and underrepresented groups. Jeffery started his organization 15 years ago because he realized men may want to help but don't know what to do on a daily basis.
>
> One of Jeffery's suggestions – go out and ask someone who is different from you about their lived experience at work. Have a "listening conversation." It may feel uncomfortable, but these conversations are critical to creating understanding and empathy.
>
> Another suggestion – download this podcast and listen to it in a team meeting; then have the conversation.
>
> Jeffery has many free tools on his website – https://lnkd.in/geMSQyqw – and I would encourage each of you to check these out! LISTEN TO THE PODCAST: https://lnkd.in/gd7if_Ft."
>
> (Jeffery Halter, speaking to Gagel, 2023 Greatness Podcast)

One friend I admire for her thoughtful social media presence is Maja Rosenquist, Senior Vice President, Mortenson Construction, as she tirelessly and positively promotes the great work of her people, her teams, and her company, and advocates for women in the industry.

Here is a great example (shared with her permission):

> "I'm really proud of this article recently published by Engineering News-Record highlighting how Mortenson has worked diligently to prioritize bringing in more women and retaining them in the industry. Just under 11% of the industry workforce and 3.7% of all tradespeople are women. Our efforts are paying off with women in our total workforce making up more than double the national rate (25%). We still have a lot to do but we're on our way to reaching parity in our industry. Read more here: https://bit.ly/3M7NHR5" (Rosenquist, 2023).

Figure 22.2 Image Maja Rosenquist Social Media Post. Source: With permission from Maja Rosenquist.

Social media is a powerful tool for helping influence the future of our industry. I encourage you to pay attention to and support the efforts of people who are utilizing social media to create change in our industry (Figure 22.2).

Activity:

- How are you engaging (or how might you engage) in social media to continue improving our industry?
- What one step might you take to more fully engage in the social media conversation of our industry?

---

9) **Find Other Ways to Contribute**

Through my work in the Australian pipeline industry, I came to know Brad Buchanan, Sales Director with Prime Creative Media. Brad is a wonderful advocate for our industry, and reached out to me when his company was starting a new magazine in Australia called *Inside Construction*. Brad asked if I would like to be an outside contributor with two pages of content in each edition, if that would be worth my time. I asked only one question: "Can I write about the things I care about?" And Brad replied, "Yes." This is one additional way I can contribute to our thinking in the industry.

My Greatness podcast and those of other leaders in our industry also contribute to creating change.

"*Let me start with the name because everyone loves the name* **You Don't Look Like an Engineer.** *I run the podcast with my friend Sohan and we chose the name because usually when we are introduced to someone in the industry the first question we are asked is,* "*What do you do for a living?*" *and we say,* "*Oh, we're in engineering,*" *and the first thing that they say is,* "*Oh, you don't look like an engineer.*" *We love putting people in boxes and tags as human beings.*

"*Sohan and I met in the industry and apart from our love of engineering we have so many other things in common like gender equity and saving communities and humanitarian engineering. One of the things that we really advocate for is minorities in the industry and because we are part of that minority group, both women of color, there were so many stories that we would share with one another. I see this platform as a powerful tool to reach audiences and to also give the opportunity to all the females and other people regardless of the gender to share their story and how they navigate challenges and how these situations that we go through are real. And instead of just continuing to provide data or statistics, to actually put a name on the stories or the numbers that we share.*"

(Laura Miranda, speaking to Gagel, 2024 Greatness Podcast)

Recording, listening to, and sharing podcasts form an example of how we can lift our voices and the voices of others in manifesting change in the industry. There are many ways we can contribute our voice and support the voices of others.

> **Activity:**
> - Are there other ways you might lend your voice to efforts to build a more inclusive industry?
> - What podcast might you start listening to in support of those working to change our industry?

Conclusion

Each of us has the power to shape the future of our industry and help lead and influence in the ways I have described. Society is counting on us to continue building and maintaining critical assets and solving complex problems. Tapping into diverse talent pools and creating a construction industry where everyone feels they belong contributes to the sustainability of our industry.

> Some might consider my induction into the National Academy of Construction and think, "Hang on, Gretchen's never held a senior position in a global construction engineering or construction firm, never achieved greatness as a construction industry academic." My induction is a reflection of my passion for contributing to systemic change in the industry through convening, advocating, and influencing. This industry has given me so much. I am committed to giving back, to shaping our future, and I encourage you to do the same.

References

Construction Industry Culture Taskforce. (n.d.). *About us: Addressing major issues impacting productivity*. Available at: https://www .constructionindustryculturetaskforce.com.au/about-us/ [Accessed: 16 June 2024].

Covey, S. R. (2020). *The 7 habits of highly effective people*. 30th anniversary edn. New York: Simon & Schuster.

Crookes, A. (2024). Interview by Gretchen Gagel [Zoom]. May 21.

Gagel, G. and Burgess, D. (2023). *Denise Burgess discusses her career as an African American woman leader in the construction industry* [Podcast]. 29 December. Available at: https://open.spotify.com/episode/77UCAi1FXUKeMOS6yfes8z [Accessed: 2 May 2024].

Gagel, G., Davies, S., and Robb, C. (2024). *Steve Davies and Cara Robb discuss the Australian Pipelines and Gas Association women's leadership development program* [Podcast]. 26 January. Available at: https://open.spotify.com/episode/0RKRSsLphJ0tWXiJqBztJg [Accessed: 2 May 2024].

Gagel, G. and Gerber, P. (2022). *Dr. Paula Gerber discusses founding the National Association of Women in Construction Australia* [Podcast]. 18 February. Available at: https://open.spotify.com/episode/5xoc4xs1pLAZ11p0CE1AAG [Accessed 29 June 2024].

Gagel, G. and Halter, J. T. (2023). *Jeffery Tobias Halter discusses how we create inclusive organizations* [Podcast]. 17 November. Available at: https://open.spotify.com/episode/5HcF8ZcTU01guNNUsb8UI2 [Accessed: 2 May 2024].

Gagel, G. and Jones, E. (2024). *Erica Jones discusses women's leadership in construction* [Podcast]. 23 February. Available at: https://open.spotify.com/episode/3RQth0SyIU9nWVsm4RSqKg [Accessed: 2 May 2024].

Gagel, G. and McCombs, B. (2023). *Brandy McCombs discusses leadership in the construction industry* [Podcast]. 20 October. Available at: https://open.spotify.com/episode/6LcqOpIQWNC362U2pokSJs [Accessed: 2 May 2024].

Gagel, G. and Miranda, L. (2024). *Laura Miranda discusses her podcast "You Don't Look Like an Engineer"* [Podcast]. 8 March. Available at: https://open.spotify.com/episode/6sLfjJDdlYzWoT5zkSD1nR [Accessed: 2 May 2024].

Gagel, G. and Trainor, G. (2021). *Gabrielle Trainor discusses the construction industry culture taskforce* [Podcast]. 7 June. Available at: https://open.spotify.com/episode/4M9xOoaFNvqPQUb7Wu54hZ [Accessed: 28 April 2024].

Redwin, M. (2021). 'How the ACA corrected the gender imbalance in the boardroom to drive positive change across the industry,' *The NAWIC Journal 25th Anniversary Edition.* Available at: https://issuu.com/nawicau/docs/the_nawic_journal_25th_anniversary_edition/s/13377604 [Accessed: 29 June 2024].

Rosenquist, M. (2023). 'Empowering women in craft and leadership roles,' *Engineering News Record,* 27 March. Available at: https://www.enr.com/articles/56159-empowering-women-in-craft-and-leadership-roles [Accessed: 16 June 2024].

Conclusion

While writing this book, I took three road trips on the Australian coast to clear my head and focus. One night, I checked into my hotel in Forster and the desk clerk suggested I walk down to the water in the evening to see the dolphins. This sounded like a great idea, so down to the water I went. For several minutes, I searched for the dolphins in the water. Nothing. I thought to myself, "Are you dense?" The hotel clerk had made it sound so easy. I started to return to the hotel, and instead walked up to a woman sitting on a bench and said, "I've never been here before and I hear there are dolphins to see." She immediately pointed to an area of water at a right angle to where we faced. Dozens of dolphins—and I took this beautiful picture.

Building Women Leaders: A Blueprint for Women Thriving in Construction,
First Edition. Gretchen Gagel.
© 2025 John Wiley & Sons, Inc. Published 2025 by John Wiley & Sons, Inc.

Asking for help is not a weakness. Had I not asked this woman for help, I would have missed the spectacular sight of these dolphins with the beautiful sunset behind them. Seeking knowledge is a sign of great strength.

I think back to my early days when my dear friend Dennis Smith and I officed next to one another. Dennis is one of the brightest people I know, and about once a year when we were miraculously in the office on the same day, Dennis would walk into my office and ask, "I have this client and here's what I'm trying to accomplish, any thoughts on techniques?" and I'd be off to the races with ideas. After about five minutes, Dennis would say "thank you" and return to his office to sort through the ideas. You see, Dennis and I had both completed the Highlands Profile, and Dennis knew I was high on "idea productivity;" not necessarily good ideas, but I could come up with many ideas in a short amount of time. Dennis was being smart in utilizing the skills of a teammate by asking for help.

I encourage you to be kind to yourself. Leadership is a journey. You will have amazing days where you think to yourself, "I've got this!" You will also have clunker days where you think, "Wow, I really messed that up." It is like golf. Every now and again, I hit a great shot and think, "I know how to play this game!" And then… well, into a bunker, into a tree, or sometimes even into the ocean! I was telling a friend and mentor about something I had done that I was embarrassed about, and she said to me, "Welcome to the human race." We all make mistakes. Learn from them. Celebrate your wins. If you aspire to be a great leader, you will never stop learning.

> Skye Mason, Director of Operations for Lendlease, shares this: "*Be true to what you believe in, be a model of what you believe in. Get the people right first and everything will follow. Emotional intelligence is critical. Be honest with yourself and where you need to improve. We work hard, almost to our disservice. We don't need to justify why we are in a role.*"
>
> (Mason, 2024)

I love this thought by friend and author Amy Su: "*I think my big message is my going in assumption is that we're all whole. We have everything we need on the inside. So, when you find yourself in a Leader B moment, I hope it is met with compassion and to just explore what's actually bothering you. Let me attend to that, let me reopen the door to my best self. That is what I most hope for everybody listening to us today*" (Amy Jen Su, speaking to Gagel, 2021 Greatness Podcast).

> "*Know your value, be sure of yourself, be okay to be the diverse you. Recognize it takes every skill set to make a project and don't give up, it's a good industry.*"
>
> (McDowall, 2023)

I appreciate you taking the time to share in my experiences and knowledge, as well as the knowledge of the amazing people I've included in the book. I am a "raving fan" of each of you—go forth and lead!

References

Gagel, G. and Su, A. J. (2021). *Amy Jen Su discusses her book The Leader You Want to Be* [Podcast]. 12 November. Available at: https://open.spotify.com/episode/4ePtBXmgN40IG7S7pKiFu6 [Accessed: 1 May 2024].

Mason, S. (2024). Interview by Gretchen Gagel [Zoom]. 1 December.

McDowall, D. (2023). Interview by Gretchen Gagel [Zoom]. 30 November.

Appendix A

Construction Industry Leaders Interviewed for This Book

Alam, Badar, Construction Competency Leader/Manager Enterprise Construction, The Chemours Company

Andresky, Christa, Executive Vice President and Chief Financial Officer, Turner Construction Company

Bainbridge, Lou, CEO, Bainbridge Consulting

Bang, Avery, Partner, Mulago Foundation

Beaudry, Jane, HSE Director Life Sciences North America, Jacobs

Blackwell, Kelly, Director, Strategic Sourcing & Procurement - Facilities, Real Estate and EOHSS, Bristol Myers Squibb

Breen, James, Vice President Pharmaceutical Global Engineering and Technology, Johnson & Johnson

Browning, Andy, General Manager of Engineering and Construction Services, Duke Energy Corporation

Crookes, Annabel, Director - Legal and Risk, Laing O'Rourke Australia; Chair, Australian Constructors Association (ACA)

De Rosi, Julia, Deputy Assistant Commissioner, Office of Project Delivery, U.S. General Services Administration

Dora, Steve, Senior Manager, Facility Projects, Toyota North America

Drott, Ernest A., P.E, PMP, Chief, Engineering and Construction Division, Corp of Engineers, Great Lakes and Ohio River Division

Ellis, Jim, CEO, Ellisian; Chair, Construction Users Roundtable (CURT)

Fabra, Maria Pilar Gomez, Human Resources Director, Australia and New Zealand, Acciona

Gallagher, Brian, Vice President, Corporate Development, Graycor

Gill, Dominique, Founder and Managing Director, Urban Core

Gosser, Maddie, HR OpX Program Manager, Baker Construction

Gu, Tracy, Vice President, Cockram Construction USA

Henderson, Kelcey, President and Co-Founder, Continuum Advisory Group

Hinz, Lisa, Founder and Owner, The Confidence Track

Building Women Leaders: A Blueprint for Women Thriving in Construction,
First Edition. Gretchen Gagel.
© 2025 John Wiley & Sons, Inc. Published 2025 by John Wiley & Sons, Inc.

Hooper-Berdik, Meaghan E., Senior Vice President New England, Turner Construction Company

Houssais, Ornella, Assistant Project Manager, Brinkman Construction

Jackson, Barbara, President, Barbara Jackson LLC

Kohn, Micki, Executive Vice President of Project and Support Services, Hargrove Engineers and Constructors

Kounkel, Courtney, Founder and Owner, Monarch Build

Magnus, Teresa, Managing Director, Magnus and Company

Mason, Skye, Director of Operations, Lendlease

McDowall, Donna, Director, RP Infrastructure

Neuscheler, Kim M., Vice President & General Manager, Turner Construction Company

Nussmeier, Bob, Vice President of Strategic Clients, Baker Construction

Petrillo, Kerri Smith, Chief Talent and Strategy Officer, Baker Construction

Ramsey, Anne, Director, Global Project, Construction, Facilities & Utilities Management, Procter & Gamble

Redwin, Meg, Executive Director, General Counsel–Global, Multiplex

Reilly, Tom, President, Turner Construction Company

Rotunno, Dominic, Senior Director, Capital Engineering, Equipment, & MRO, Bristol Myers Squibb

Salam-Haughton, Sharmeena A., Branch Chief, Construction Management, U.S. Department of State Bureau of Overseas Buildings Operations

Soukup, Beth, Chief Financial Officer, JE Dunn Construction

Stagner, Laura, FAIA, DBIA, PMP, Retired Assistant Commissioner, Office of Project Delivery, U.S. General Services Administration

Sukkar, Josephine, Founder and Principal, Buildcorp

Whitney, Jayne, Chief Strategy Officer, John Holland Group

Appendix B

Greatness Podcast Guests Included in This Book

Arkle, Suzanne - President & CEO, ZANN Inc.

Arnold, Carrie - PhD, MCC, Principal Coach and Consultant, The Willow Group

Beer, Michael - Cahners-Rabb Professor of Business Administration, Emeritus, Harvard Business School

Belz, Christy - President, Coach & Consultant, Empowerment Coaching and Consulting

Blesing, Amanda - C-suite Mentor, CEO & Founder, The She-Suite® Club

Boyar, Bill - Founding Shareholder, BoyarMiller

Brower, Tracy - PhD, Sociologist, and Author

Burgess, Denise - President & CEO, Burgess Services

Chaskalson, Michael - Author and Founder, Mindfulness Works Ltd.

Chermack, Thomas J. - President, Chermack Scenarios and Professor, Organizational Learning, Performance and Change, Colorado State University

Clark, Sam - CEO, Clark Construction

Darnell, Brent - Owner, Brent Darnell International

Davies, Steve - CEO, The Australian Pipelines and Gas Association

Dolan, Gabrielle - Speaker, Podcaster, and Author; Founder of Jargon Free Fridays

Edmondson, Amy - Novartis Professor of Leadership and Management, Harvard Business School

Garner, Janine - Business Mentor, Author, and Keynote Speaker

Gerber, Paula - Law Professor and Associate Dean (Education), Monash University

Gino, Francesca - Professor of Business Administration, Harvard Business School

Goldsmith, David - Founder, The Goldsmith Group

Grayson Riegel, Deborah - CEO, Deborah Grayson Riegel Coaching and Consulting

Halter, Jeffery Tobias - President, YWomen

Hansen, Kristen - CEO, EnHansen Performance

Hedva, Beth, PhD - Clinical Psychologist and Co-Chair, Canadian Institute for Transpersonal and Integrative Sciences

Building Women Leaders: A Blueprint for Women Thriving in Construction,
First Edition. Gretchen Gagel.
© 2025 John Wiley & Sons, Inc. Published 2025 by John Wiley & Sons, Inc.

Hinz, Lisa - Founder and Owner, The Confidence Track

Jones, Erica - Vice President of Marketing, Cerris (formerly MMC Corp.)

Jourdan, Lee - Retired Chief Diversity Officer, Chevron, and Director

Mahlab, Bobbi - Founder & Chair of Mahlab, Co-Founder of Mentor Walks and Fellow Advanced Leadership Initiative, Harvard University

Marshall, Robert J - Director, Knowledge Teams International

McCombs, Brandy - President, IBC

McGeorge, Donna - Keynote Speaker, Trainer, and Coach

Meares, Anna - OAM, OLY, Decorated Australian Short Track Cycling Olympian, Speaker, and Chef de Mission, Australian Olympic Committee

Meyer, Pamela - Ph.D., President, Meyer Agile Innovation

Miranda, Laura - Podcast Host, You Don't Look Like an Engineer, Marketing Assistant, BG&E

Nawaz, Sabina - CEO, Coach, and Author

Nelson, Brendan - AO, Iconic Australian Politician and Senior Vice President of The Boeing Company and President of Boeing Global

Noonan, Andre - Chief Operating Officer, Acciona Australia & New Zealand

Pillot, Christelle - Life and Career Coach, Founder of Freedom Catcher Academy

Platow, Michael - Professor of Psychology, Australian National University (ANU) School of Medicine and Psychology

Ransom, Holly - Author, Speaker, and CEO of Emergent

Robb, Cara - Land Access Officer, CNC Management

Rosenquist, Maja - Senior Vice President, Mortenson Construction

Ross, Howard - Founder, Udarta Consulting LLC

Sales, Michelle - Director, Michelle Sales Leadership

Sandler, Amy - Principal Coach & Podcast Host, Radical Candor

Schein, Edgar - PhD (Deceased), Author, Consultant, and Professor Emeritus MIT

Schein, Peter - Co-Founder & COO, The Organizational Culture and Leadership Institute

Schwarztberg, Joel - Senior Presentation and Media Coach, Throughline Group LLC

Shamir, Alison - Imposter Syndrome Expert & International Speaker

Staun, Mike - Retired Associate Director of Global Capital Management, Procter & Gamble

Su, Amy Jen - HBR Press Book Author and Executive Advisor/Coach

Trainor, Gabrielle - AO, Chair, Construction Industry Culture Taskforce, Director

Worley, Chris - Author and Research Professor of Management, Pepperdine Graziadio Business School

Index